T0255512

Lecture Notes in Artificial Intelligence 12591

Subseries of Lecture Notes in Computer Science

More information about this subseries at http://www.springer.com/series/1244

Valerio Bitetta · Ilaria Bordino ·
Andrea Ferretti · Francesco Gullo ·
Giovanni Ponti · Lorenzo Severini (Eds.)

Mining Data for Financial Applications

5th ECML PKDD Workshop, MIDAS 2020
Ghent, Belgium, September 18, 2020
Revised Selected Papers

 Springer

Editors
Valerio Bitetta
UniCredit
Milan, Italy

Ilaria Bordino
UniCredit
Rome, Italy

Andrea Ferretti
UniCredit
Milan, Italy

Francesco Gullo
UniCredit
Rome, Italy

Giovanni Ponti
ENEA Portici Research Center
Portici, Italy

Lorenzo Severini
UniCredit
Rome, Italy

ISSN 0302-9743   ISSN 1611-3349   (electronic)
Lecture Notes in Artificial Intelligence
ISBN 978-3-030-66980-5   ISBN 978-3-030-66981-2   (eBook)
https://doi.org/10.1007/978-3-030-66981-2

LNCS Sublibrary: SL7 – Artificial Intelligence

This Springer imprint is published by the registered company Springer Nature Switzerland AG
The registered company address is: Gewerbestrasse 11, 6330 Cham, Switzerland

Preface

MIDAS 2020: The 5th Workshop on MIning DAta for financial applicationS

Workshop Description

Motivation. Like the famous King Midas, popularly remembered in Greek mythology for his ability to turn everything he touched with his hand into gold, the wealth of data generated by modern technologies, with the widespread presence of computers, users and media connected by the Internet, is a goldmine for tackling a variety of problems in the financial domain.

Nowadays, people's interactions with technological systems provide us with gargantuan amounts of data documenting collective behavior in a previously unimaginable fashion. Recent research has shown that by properly modeling and analyzing these massive datasets, for instance representing them as network structures, it is possible to gain useful insights into the evolution of the systems considered (i.e., trading, disease spreading, political elections). Investigating the impact on financial decisions of data arising from today's application domains is of paramount importance. Knowledge extracted from data can help gather critical information for trading decisions, reveal early signs of impactful events (such as stock market moves), or anticipate catastrophic events (e.g., financial crises) that result from a combination of actions, and affect humans worldwide.

The importance of data-mining tasks in the financial domain has long been recognized. For example, in the Web context, changes in the frequency with which users browse news or look for certain terms on search engines have been correlated with product trends, level of activity in certain given industries, unemployment rates, or car and home sales, as well as stock-market trade volumes and price movements. Other core applications include forecasting the stock market, predicting bank bankruptcies, understanding and managing financial risk, trading futures, credit rating, loan management, and bank customer profiling. Despite its well-recognized relevance and some recent related efforts, data mining in finance is still not stably part of the main stream of data-mining conferences. This makes the topic particularly appealing for a workshop proposal, whose small, interactive, and possibly interdisciplinary context provides a unique opportunity to advance research in a stimulating but still quite unexplored field.

Objectives and Topics. The aim of the 5th Workshop on MIning DAta for financial applicationS (MIDAS 2020), held in conjunction with the 2020 European Conference on Machine Learning and Principles and Practice of Knowledge Discovery in Databases (ECML-PKDD 2020), Virtual Conference, September 14–18, 2020, was to discuss challenges, potentialities, and applications of leveraging data-mining tasks to tackle problems in the financial domain. The workshop provided a premier forum for sharing findings, knowledge, insights, experience, and lessons learned from mining data generated in various domains. The intrinsically interdisciplinary nature of the workshop promoted interaction between computer scientists, physicists, mathematicians, economists, and financial analysts, thus paving the way for an exciting and stimulating environment involving researchers and practitioners from different areas.

Topics of interest included, among others: forecasting the stock market, trading models, discovering market trends, predictive analytics for financial services, network analytics in finance, planning investment strategies, portfolio management, understanding and managing financial risk, customer/investor profiling, identifying expert investors, financial modeling, measures of success in forecasting, anomaly detection in financial data, fraud detection, discovering patterns and correlations in financial data, text mining and NLP for financial applications, financial network analysis, time series analysis, and pitfall identification.

Outcomes. MIDAS 2020 was structured as a full-day workshop. Owing to the Covid-19 pandemic, MIDAS 2020 was organized as a fully fledged virtual event. In particular, the workshop followed a "live" mode, where presentations happened in real time, with organizers, speakers, and attendees remotely joining the event.

We encouraged submissions of regular papers (long or short), and extended abstracts. Regular papers could be up to 15 pages (long papers) or 8 pages (short papers), and reported on novel, unpublished work that might not be mature enough for a conference or journal submission. Extended abstracts could be up to 5 pages long, and presented work in progress recently published work fitting the workshop topics, or position papers. All submitted papers were peer reviewed by three reviewers from the program committee, and a selection was made on the basis of these reviews. MIDAS 2020 received 15 submissions, among which 11 papers were accepted (8 long papers and 3 short papers).

In accordance with the reviewers' scores and comments, the paper entitled "Financial Fraud Detection with Improved Neural Arithmetic Logic Units," authored by Daniel Schloer, Markus Ring, Anna Krause, and Andreas Hotho, was selected as the best paper of the workshop.

The program of the workshop was enriched by an invited speaker: Dr. Luigi Bellomarini from Banca d'Italia, who gave a talk titled "Neither in the Programs Nor in the Data: Mining the Hidden Financial Knowledge with Knowledge Graphs and Reasoning."

September 2020

Valerio Bitetta
Ilaria Bordino
Andrea Ferretti
Francesco Gullo
Giovanni Ponti
Lorenzo Severini

Organization

Program Chairs

Valerio Bitetta	UniCredit, Italy
Ilaria Bordino	UniCredit, Italy
Andrea Ferretti	UniCredit, Italy
Francesco Gullo	UniCredit, Italy
Giovanni Ponti	ENEA, Italy
Lorenzo Severini	UniCredit, Italy

Program Committee

Aris Anagnostopoulos	Sapienza University of Rome, Italy
Argimiro Arratia	Universitat Politécnica de Catalunya, Spain
Antonia Azzini	C2T, Italy
Xiao Bai	Yahoo Research, USA
Luca Barbaglia	JRC - European Commission, Italy
Ludovico Boratto	Eurecat, Spain
Cristian Bravo	Western University, Canada
Doug Burdick	IBM Research, USA
Alejandra Cabana	Universitat Autònoma de Barcelona, Spain
Matteo Catena	NTENT, Spain
Jeremy Charlier	National Bank of Canada, Canada
Sergio Consoli	JRC - European Commission, Italy
Jochen De Weerdt	KU Leuven, Belgium
Carlotta Domeniconi	George Mason University, USA
Wouter Duivesteijn	Eindhoven University of Technology, The Netherlands
Amita Gajewar	Microsoft, USA
Edoardo Galimberti	Independent researcher, Italy
Gergely Ganics	Corvinus University of Budapest, Hungary
Cuneyt Gurcan Akcora	University of Manitoba, Canada
Roberto Interdonato	CIRAD, France
Rajasekar Krishnamurthy	IBM Almaden, USA
Yelena Mejova	ISI Foundation, Italy
Sandra Mitrovic	KU Leuven, Belgium
Davide Mottin	Aarhus University, Denmark
Alan Perotti	ISI Foundation, Italy
Luca Rossini	Vrije Universiteit Amsterdam, The Netherlands
Dongjin Song	NEC Laboratories America, USA
Letizia Tanca	Politecnico di Milan, Italy

Contents

Trade Selection with Supervised Learning and Optimal Coordinate Ascent (OCA)

David Saltiel[1,2(✉)], Eric Benhamou[2,3(✉)], Rida Laraki[3(✉)], and Jamal Atif[3(✉)]

[1] LISIC - Universite du Littoral - Cote d'Opale, Calais, France
[2] A.I. SQUARE CONNECT, 35 bd. d'Inkermann, 92200 Neuilly sur Seine, France
{david.saltiel,eric.benhamou}@aisquareconnect.com
[3] LAMSADE, Universite Paris Dauphine, 75016 Paris, France
{rida.laraki,jamal.atif}@dauphine.fr

Abstract. Can we dynamically extract some information and strong relationship between some financial features in order to select some financial trades over time? Despite the advent of representation learning and end-to-end approaches, mainly through deep learning, feature selection remains a key point in many machine learning scenarios. This paper introduces a new theoretically motivated method for feature selection. The approach that fits within the family of embedded methods, casts the feature selection conundrum as a coordinate ascent optimization with variables dependencies materialized by block variables. Thanks to a limited number of iterations, it proves efficiency for gradient boosting methods, implemented with XGBoost. In case of convex and smooth functions, we are able to prove that the convergence rate is polynomial in terms of the dimension of the full features set. We provide comparisons with state of the art methods, Recursive Feature Elimination and Binary Coordinate Ascent and show that this method is competitive when selecting some financial trades.

1 Introduction

Feature selection is also known as variable or attribute selection. This method concerns the selection of a subset of relevant attributes in our data that are most relevant to our predictive modeling problem. It has been an active and fruitful field of research and development for decades in statistical learning. It has proven to be effective and useful in both theory and practice for many reasons: enhanced learning efficiency and increasing predictive accuracy (see [22]), model simplification to ease its interpretation and improve performance (see [1,19] and [5]), shorter training time (see [22]), curse of dimensionality avoidance, enhanced generalization with reduced overfitting, and implied variance reduction. Both [15] and [14] are nice references to get an overview of various methods to tackle feature selections. The approaches followed vary; briefly speaking, the methods can be sorted into three main categories: Filter method, Wrapper methods and Embedded methods. We develop these three categories in the following section.

© Springer Nature Switzerland AG 2021
V. Bitetta et al. (Eds.): MIDAS 2020, LNAI 12591, pp. 1–15, 2021.
https://doi.org/10.1007/978-3-030-66981-2_1

1.1 Filter Methods

Filter type methods select variables despite the model. These methods suppress the least interesting variables using ranking techniques as a criteria to select the variables. Once the ranking is done, a threshold is determined in order to select features with rank above it. These methods are very effective in terms of computation time and robust to overfitting. One famous Filter method approach is the algorithm developed in [16] in 1992, called *Relief*, for application to binary classification problems. By construction, Filter methods may select redundant variables as they do not consider the relationships between variables. One of the most used criteria for Filter methods is the Pearson correlation coefficient, which is simply the ratio between the covariance and the square root of the two variances: $\mathrm{Cov}(x_i, y)/\sqrt{\mathrm{Var}(x_i)\,\mathrm{Var}(y)}$ with x_i the ith feature in the model and y the label associated. It is well known that this correlation ranking can only detect linear dependencies between features ant the target label. The Filter method procedure is summarized in Fig. 1.

Fig. 1. Filter method: it consists in 4 specific steps. Arrows emphasize that these steps are done in chronological order.

1.2 Wrapper Methods

Wrapper methods allow detecting possible interactions between variables by evaluating subsets of them. In Wrapper methods, a model must be trained to test any subsequent feature subset. Consequently, these methods are iterative and computationally expensive. However, these methods can identify the best performing features set for that specific modeling algorithm. Theses methods are explained in Fig. 2. Some known examples of Wrapper methods are forward and backward feature selection methods. These methods are presented in the work of [18]. The backward elimination starts with all features and progressively remove them. At the opposite, the forward selection starts with an empty set and progressively add them. If we have n features, we need to train n classifiers for the

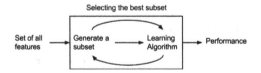

Fig. 2. Wrapper method: like filter method, it consists in 4 different steps but there are iterations between steps 2 and 3 emphasized by the rectangle until each feature subset combination is computed.

first step, then $n - 1$ classifiers for the second step and so on. We then have $n(n + 1)/2$ training steps for both methods. However, forward selection starts with small features subsets so it can be computationally cheaper if the stopping condition is satisfied early. An adaptive forward-backward algorithm is proposed in [31]. One of the state of the art Wrapper method is Recursive Feature Elimination (RFE) (see for instance [21] for more details). It first fits a model and removes features until a pre-determined number of features. Features are ranked through an external model that assigns weights to each features and RFE recursively eliminates features with the least weight at each iteration. One of the main limitation to RFE is that it requires the number of features to keep. This is hard to guess a priori and one may need to iterate much more than the desired number of feature to find an optimal feature set.

1.3 Embedded Methods

Embedded methods perform feature selection as a part of the modeling algorithm's execution. Many hybrid methods are developed to combine the advantages of Wrapper and Filter methods. These methods combine learning and feature selection through a prior or a posterior regularization. The different methods of regularization are given in [11]. The use of L1 regularization is introduced in [26] and a comparison between L1 and L2 regularization is explained in [24]. Besides, an extension to non-linear spaces is given in [3], using Multiple Kernel Learning (MLKL). These methods are summarized in Fig. 3. Very recently there are also many methods brought by the field of machine learning interpretability, like Shapley values calculation [25] and local surrogates (LIME) [2] that can be used for features selection being run on the train set.

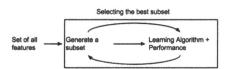

Fig. 3. Embedded method: as opposed to filter and wrapper methods, there are only 3 steps as the learning and performance steps are combined into a single step. Like for the wrapper method, we iterate between step 2 and 3 until we find the best features subset among all combinations.

2   Framework

2.1   Block Variables

[9] deals with the choice of the number of block to consider in a feature selection problem. We propose an answer to the open problem discussed in [4] about solving an unconstrained minimization using a projection method in which each iteration consists of performing a gradient projection step with respect to a

certain block taken in a cyclic order or using an alternating minimization method using only two blocks. Indeed, we decide to split the data into $n \geq 2$ blocks $\left([B^1] \ldots [B^n] \right)$ and instead of simply performing a gradient ascent method on a single block at each iteration, we first initialize our method by finding the best sub-block data set in terms of prediction score for our supervised learning problem and then apply a gradient ascent method on each single block at every iteration (for more details, see Sect. 3).

2.2   Result of Convergence

In order to motivate our method that relies on coordinate ascent, we recall some theoretical results about the convergence of coordinate ascent optimization. The theory is well understood for the convex case (see [29]). The non convex case without gradient which is our example is however much harder as we have local minima issue and mathematical assumptions too weak to be able to prove convergence. However, convergence results under strong convex conditions provide some hint about the efficiency of this method and its convergence rate that is linear. The proof is given by [23]. In order to have some meaningful result, we need to make some necessary assumptions for our function f to be minimized. We assume that we have a real value n-dimensional function $f : \mathbb{R}^n \to \mathbb{R}$. In this section, we stick to the traditional presentation and examine minimization to make proof reading easier. We examine the following optimization program:

$$\min_{x \in \mathbb{R}^n} f(x) \tag{1}$$

The notation and assumptions used for the two following proposition are given by [23].

Proposition 1. *Under assumption given by [23], coordinate ascent optimization converges to the global minimum f^* at a linear rate proportional to $2nL_{\max}R_0^2$, that is*

$$\mathbb{E}[f(x_k)] - f^\star \leq \frac{2nL_{\max}R_0^2}{k} \tag{2}$$

Proof. The proof is given by [23].

We have an additional control of our convergence rate which is given by the following proposition.

Proposition 2. *Under assumption [23], and for $\sigma > 0$, coordinate ascent optimization error is controlled by the following inequality*

$$\mathbb{E}[f(x_k)] - f^\star \leq \left(1 - \frac{\sigma}{nL_{\max}} \right)^k (f(x_0) - f^\star) \tag{3}$$

Proof. The proof is given by [23].

3   Method Developed

In many applications, we can regroup features among families. We call these features block variables. Typical example is to regroup variables that are observations of some physical quantity but at a different time (like the speed of the wind measure at different hours for some energy prediction problem, the price of a stock in an algorithmic trading strategy for financial markets, the temperature or heart beat of a patient at different time, ...). We denote from now $\mathcal{M}_{(n,p)}$ the space of matrices of n rows and p columns. Supposing that we have J rows of data, we can formally regroup our variables into two sets:

- the first set encompasses $[B^1]\ldots[B^n]$. These are block variables of different length L_i. Mathematically, the sub-block variables are denoted by B^i_j with $(B^i_j)_{\substack{i\in 1\ldots n \\ j\in 1\ldots L_i}}$ taking value in $\mathbb{R}^J$.
- the second set is denoted $[S]$ and is a block of p single variables.

Graphically, our variables looks like that:

$$
\begin{pmatrix}
\overbrace{B^1_1 \ldots\ldots B^1_{L_1}}^{B^1} & \overbrace{B^n_1 \ldots\ldots B^n_{L_n}}^{B^n} & \overbrace{S_1 \ldots\ldots S_p}^{S} \\
\bullet \ldots\ldots \bullet & \bullet \ldots\ldots \bullet & \bullet \ldots\ldots \bullet \\
\vdots \qquad \vdots & \vdots \qquad \vdots & \vdots \qquad \vdots \\
\bullet \ldots\ldots \bullet & \bullet \ldots\ldots \bullet & \bullet \ldots\ldots \bullet
\end{pmatrix}
$$

Thus, we have N variables split between block variables and single variables, hence $N = N_B + p$ with $N_B = \sum_{i=1}^{n} L_i$.

Our algorithm works as follows. We first fit our classification model to find a ranking of features importance. The performance is computed using the Gini index for each variable. Gini Index, or Gini impurity, computes the probability of a specific feature that is classified incorrectly when selected randomly. It can be mathematically written as: $Gini = 1 - \sum_{i=1}^{n} P_i^2$ with P_i the probability of a feature being classified for some distinct class. We then keep the first k best ranked features for each blocks $B^1 \ldots B^n$ in order to find the best initial guess for our coordinate ascent algorithm. Notice that the set of unique variables is not modified during the first step of the procedure. The objective function is the number of correctly classified samples at each iteration. It can be expressed as followed:

$$
accuracy = \frac{TP + TN}{\text{number of examples}} \tag{4}
$$

with True Positives (TP) denoting the number of positive examples labeled as positive and True Negatives (TN) the number of negative examples labeled as negative.

Remark 1. Notice that the number of examples (denoted by n in this remark) respects the following relation : $n = TP + TN + FP + FN$ with False Positives (FP) denoting the number of negative examples labeled as positive and False Negatives (FN) the number of positive examples labeled as negative.

We then enter the main loop of the algorithm. Starting with the vector of $(k, \ldots, k, \ \mathbb{1}_p^T)$ as the initial guess for our algorithm, we perform our coordinate ascent optimization in order to find the set with optimal score and the minimum number of features. The coordinate ascent loop stops whenever we either reach the maximum number of iterations or the current optimal solution has not moved so far between two steps.

Taking the previous notation, we start with the initial block data set $\left(\left[B_1^1 \ldots B_{L_1}^1 \right] \ldots \left[B_1^n \ldots B_{L_n}^n \right], [S] \right)$ and reduce it to the new data set $\left(\left[B^{1, \Diamond} \right] \ldots \left[B^{n, \Diamond} \right], [S] \right)$, with $B^{i, \Diamond} \in \mathcal{M}_{(J, k)} \quad \forall i \in 1 \ldots n$. We can first apply a coordinate ascent optimization on each single block and get a new data set structure without considering blocks. We secondly apply a binary coordinate ascent optimization on the remain data set in order to get the final data set $\left(\left[B^{1, \star} \right] \quad \ldots \quad \left[B^{n, \star} \right], \ [S^{\star}] \right)$. We summarize the algorithm in the pseudo code 1. To control early stopping, we use precision variables denoted by ε_1 and ε_2, and two maximum iterations Iteration max_1 and Iteration max_2 that are initialized before starting the algorithm. We also denote $\text{Score}(k_1, \ldots, k_n, \mathbb{1}_p)$ to be the accuracy score of our classifier with each B_i block of variables retaining k_i best variables and with single variable all retained.

Remark 2. The originality of this coordinate ascent optimization is to regroup variable by blocks, hence it reduces the number of iterations compared to Binary Coordinate Ascent (BCA) as presented in [30]. The stopping condition can be changed to accommodate for other stopping conditions.

Remark 3. The specificity of our method is to keep the j best representative features for each features class, as opposed to other methods that only select one representative feature from each group, ignoring the strong similarities between each feature of a given variable block. This takes in particular the opposite view of feature Selection with Ensembles, Artificial Variables, and Redundancy Elimination as developed in [27].

4 Numerical Results

4.1 Data Set

We carry out our experiment on real finance data set. In financial markets, algorithmic trading has become more and more standard over the last few years. The rise of the machine has been particularly significant in liquid and electronic markets such as foreign exchange and futures markets reaching between 60 to 80% of total traded volume (see for instance [7,13] or [6] for more details on the various markets). These strategies are even more concentrated whenever there are very fast market moves as reported in [17]. A trading strategy is usually defined with some signal that generates a trading entry. But once we are in position, then main following point is the trading exit strategy. There are multiple method to

---

Algorithm 1. OCA algorithm :

J Best optimization

We retrieve features importance from a fitted model

We find the index $k^\star$ that gives the best score for variables block of same size k:

$$k^\star \in \underset{k \in \mathbb{R}^{L_{\min}}}{\operatorname{argmax}} \operatorname{Score}(k, \ldots, k, \mathbb{1}_p)$$

Initial guess : $x^0 = (k^\star, \ldots, k^\star, \mathbb{1}_p)$

while $|\operatorname{Score}(x^i) - \operatorname{Score}(x^{i-1})| \geq \varepsilon_1$ and $i \leq$ Iteration max$_1$ **do**

$\quad x_1^i \in \underset{j \in \mathbb{R}^{L_1}}{\operatorname{argmax}} \operatorname{Score}(j, \ x_2^{i-1}, \ x_3^{i-1}, \ \ldots, \ x_n^{i-1}, \mathbb{1}_p)$

$\quad \ldots$

$\quad x_n^i \in \underset{j \in \mathbb{R}^{L_n}}{\operatorname{argmax}} \operatorname{Score}(x_1^i, \ x_2^i, \ x_3^i, \ \ldots, j, \mathbb{1}_p)$

$\quad$ i $+= 1$

end while

Full coordinate ascent optimization

Use previous solutions: $X^* = (x_1^i, \ldots, x_n^i, \mathbb{1}_p)$

$Y^* = \operatorname{Score}(X^*)$

while $|Y - Y^*| \geq \varepsilon_2$ and iteration $\leq$ Iteration max$_2$ **do**

$\quad$ **for** i=1 $\ldots$ N **do**

$\quad\quad X = X^*$

$\quad\quad X_i = \operatorname{not}(X_i^*)$

$\quad\quad$ **if** $\operatorname{Score}(X) \geq \operatorname{Score}(X^*)$ **then**

$\quad\quad\quad X^* = X$

$\quad\quad$ **end if**

$\quad$ **end for**

$\quad Y = \operatorname{Score}(X^*)$

$\quad$ iteration $+= 1$

end while

Return X^*, Y^*

---

handle efficient exits, ranging from fixed target and stop loss, to dynamic target and stop loss. Indeed, to enforce success and crystallize gain or limit loss, a common practice is to associate to the strategy a profit target and stop loss as described in various papers ([10,12,20], or [28]). We use our algorithm to do a supervised classification according to some a priori features. We are given 1500 trades over a ten years history with 135 features that can be classified into five blocks of twenty variables, one block of thirty variables and five single variables. Each block can be interpreted as a temporally relationship between some technical analysis indicators or price history. We know for each trade whether it is a 'good' or 'bad' trade (thanks to the gain and loss, denoted by PnL, engendered by the trade). The idea is to use the minimum number of features to classify a priori this data set. We use cross validation with 70% for the training set and 30% for the test sets. The procedure is summarized in Fig. 4. For full reproducibility, full data set and corresponding python code for this algorithm is available publicly on github: https://github.com/aisquareconnect/features_selection_oca. We consider a gradient boosting framework by using XGBoost library, initially

Fig. 4. Learning process for our trade selection challenge. We first use a proprietary trading strategy that generates some samples trades. We take various measures before the trades are executed to create a feature set. We combined these to create a supervised learning classification problem. Using xgboost method and OCA, we learn model parameters on a train set. We monitor overall performance of the trading strategy on a separate test set to validate scarce overfitting.

developed in [8], in the implementation of our algorithm. It is illuminating to look at the histogram of gain and losses of our trades over our 10 years of history. We can observe two peaks corresponding to the profit target and stop loss level as shown in Fig. 5. It is much better to use the gain and loss curve in the native currency of the underlying instrument than to look at the consolidated currency of our trading strategies to avoid foreign exchange noise. We will consider two data sets of a 'bad' strategy, called strategy 1, and a 'good' one, called strategy 2; each strategy engendering trades among a fixed period of time.

The graph of the evolution of the gain and loss of the two strategies is given in Fig. 6.

Remark 4. The data set of the strategy 1 is quite balanced in terms of label (PnL engendered by the strategy) whereas it is not the case for the second one, we can't thus apply the definition of the accuracy given in (4) because it would perform badly and it is recommended to use the definition of the balance accuracy, that is (taking the previous notation introduced in (4) and Remark 1:

$$\text{Balance accuracy} = \frac{TPR + TNR}{2}$$

with the True Positive Rate (TPR) and the True Negative Rate (TNR) defined as :

$$TPR = \frac{TP}{TP + FN}, \quad TNR = \frac{TN}{TN + FP}$$

True Positive Rate (TPR) concentrates on getting accurate results on the positive labels, regardless of false results. On the opposite, True Negative Rate (TNR) focuses on avoiding to incorrectly classify as a true label something that is a negative label. This is similar to type I and type II error in statistical tests.

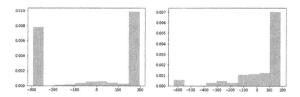

Fig. 5. Histogram of the PnL of the two strategies in native currency. The left histogram corresponds to the strategy 1 and the right one to the second one. As the algorithm is performed on an underlying asset listed on an American financial market, the resulting strategy is denominated in USD. The strategy uses fixed stop loss and profit target level. This leads to two major columns corresponding for the left peak to trades that end in a loss when they hit the stop loss level and for the right peak to trades that exit with a profit as they reach the profit target.

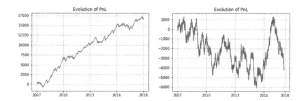

Fig. 6. Evolution of the gain and loss of the two strategies.

4.2   Comparison

We first analyze the results given by the strategy 1. We compare our method to two other methods that are supposed to be state of the art for feature selections, namely RFE and BCA. Our new method achieves a score of 62.80% with 16% of features used, to be compared to RFE that achieves 62.80% with 19% of features used. BCA performs poorly with a highest score given by 62.19% with 27.08% of features used. If we take in terms of efficiency criterium, the highest score with the less feature, our method is the most efficient among these three methods. In comparison, with the same number of features, namely 16.6%, RFE gets a score of 62.39%. All these figures are summarized in the Table 1.

5   Discussion

Compared to BCA our method reduces the number of iterations as it uses the fact that variables can be regrouped into categories or classes. The number of iterations for both OCA and BCA methods is provided below in Fig. 7. Our method requires only 350 iterations steps ton converge as opposed to BCA that needs up to 700 iterations steps as it considers blindly variables ignoring similarities between them.

Table 1. Method comparison for the strategy 1: for each row, we provide in red the best(s) (hotest) method(s) and in blue the worst (coldest) method, while intermediate methods are in orange. We can notice that OCA achieves the higher score with the minimum feature sets. For the same cardinal of features set, RFE performs worst or equally, if we want the same performance for RFE, we need to have a larger feature set. BCA is the worst method both in terms of score and minimum feature set.

Method	% of features	Score (in %)
OCA using 24 features	16.6	62.8
RFE using 24 features	16.6	62.39
BCA using 39 features	27.08	62.19
RFE using 28 features	19.4	62.8

Table 2. Method Comparison for the strategy 2:

Method	% of features	Score (in %)
OCA using 10 features	6.75	74.28
RFE using 10 features	6.75	50.47
BCA using 52 features	35.13	77.14

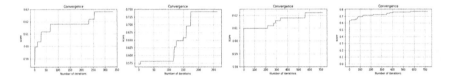

Fig. 7. Iterations steps up to convergence for OCA and BCA. OCA method is above while BCA is below. The figures on the left represent the strategy 1 and the strategy 2 is represented on the right. We see that OCA requires around 250 or 350 iteration steps to converge whereas BCA requires the double (around 700 iteration) steps to converge.

Graphically, we can determine the best candidates for the three methods listed in Table 2 in Figs. 8, 9 and 10 for both strategies. We have taken the following color code. The hottest (or best performing) method is plotted in red, while the worst in blue. Average performing methods are plotted in orange. In order to compare finely OCA and RFE, we have plotted in Fig. 9 the result of RFE for used features set percentage from 10 to 30%. We can notice that for the same feature set as OCA, RFE has a lower score and equally that to get the same score as OCA, RFE needs a large features set.

5.1   Reduced Overfitting

We look at the final goal which is to compare the trading strategy with and without machine learning. A standard way in machine learning is to split our data set between a randomized training and test set. We keep one third of our

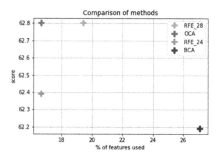

Fig. 8. Comparison between the 3 methods for strategy 1. To qualify the best method, it should be in the upper left corner. The desirable feature is to have as little features as possible and the highest score. We can see that the red cross that represents OCA is the best. The color code has been designed to ease readability. Red is the best, orange is a slightly lower performance while blue is the worst. (Color figure online)

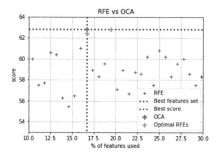

Fig. 9. Comparison between OCA and RFE for strategy 1: For RFE, we provide the score for various features set in blue. The two best RFE performers points are the orange cross marker points that are precisely the one listed in table 1. The red cross marker point represents OCA. It achieves the best efficiency as it has the highest score and the smallest feature set for this score. (Color figure online)

data for testing to spot any potential overfitting. If we use the standard and somehow naive way to take randomly one third of the data for our test set, we break the time dependency of our data. This has two consequences. We use in our training set some data that are after our test sets which is not realistic compared to real life. We also neglect any regime change in our data by mixing data that are not from the same period of time. However, we can do the test on this mainstream approach and compare the trading strategy with and without machine learning filtering. This is provided in Fig. 11. The orange (respectively blue) curve represents our algorithmic trading strategy without any machine learning filtering (respectively with filtering done by the xgboost method trained with OCA). Since the blue curve is above the orange one, we experimentally validate that using machine learning enhances the overall profitability of our trading strategy by avoiding the bad trades.

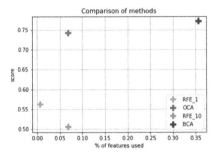

Fig. 10. Comparison between the 3 methods for strategy 2. We keep the same color code as before. The best performer point for the RFE method has a less percentage of features used than the OCA method but the score of the last one is much better. Besides, the BCA method leads to a higher score than the two other methods but with a larger features set. (Color figure online)

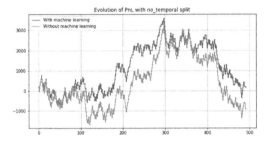

Fig. 11. Evolution of the PnL for strategy 1 with a randomized test set. The orange curve represents our algorithmic trading strategy without any machine learning filtering while the blue line is the result of the combination of our algorithmic trading strategy and the oca method to train our xgboost method. (Color figure online)

If instead we split our set into two sets that are continuous in time, meaning we use as a training test the first two third of the data when there are sorted in time and as a test set the last third of the data, we get better result as the divergence between the blue and orange curve is larger. An explanation of this better efficiency may come from the fact that the non randomization of the training set makes the learning for our model easier and leads to less overfitting overall. This method of splitting the two sets is illustrated in Figs. 12 and 13. Since the blue curve is above the orange one for both strategies, we experimentally validate that using machine learning enhances the overall profitability of our trading strategy by avoiding the bad trades.

Fig. 12. Evolution of the PnL for strategy 1 with a test set given by the last third of the data to take into account temporality in our data set. The orange curve represents our algorithmic trading strategy without any machine learning filtering while the blue line is the result of the combination of our algorithmic trading strategy and the OCA method to train our xgboost method. (Color figure online)

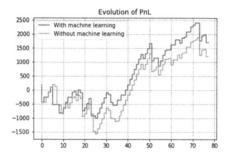

Fig. 13. Evolution of the PnL for strategy 2 with temporal split. We keep the same color code as before. (Color figure online)

6   Conclusion

In this paper, we have presented a new method, called Optimal Coordinate Ascent (OCA) that allows us selecting features among block and individual features. OCA relies on coordinate ascent to find an optimal solution for gradient boosting methods score (number of correctly classified samples for our data). OCA takes into account the notion of dependencies between variables forming blocks in our optimization. The coordinate ascent optimization solves the issue of the NP hard original problem where the number of combinations rapidly explodes making a grid search unfeasible. It transforms the NP hard problem of finding the best features into a polynomial search one. Comparing result with two other methods, Binary Coordinate Ascent (BCA) and Recursive Feature Elimination (RFE), we find that OCA leads to the minimum feature set with the highest score. Hence, OCA provides empirically the most compact data set with optimal performance.

References

1. Almuallim, H., Dietterich, T.G.: Learning Boolean concepts in the presence of many irrelevant features. Artif. Intell. **69**, 279–305 (1994)
2. Alvarez-Melis, D., Jaakkola, T.S.: On the robustness of interpretability methods. CoRR (2018). http://arxiv.org/abs/1806.08049
3. Bach, F.R.: Exploring large feature spaces with hierarchical multiple kernel learning. In: Koller, D., Schuurmans, D., Bengio, Y., Bottou, L. (eds.) Advances in Neural Information Processing Systems 21, pp. 105–112 (2009)
4. Beck, A., Tetruashvili, L.: On the convergence of block coordinate descent type methods. SIAM J. Optim. **23**(4), 2037–2060 (2013)
5. Blum, A.L., Langley, P.: Selection of relevant features and examples in machine learning. Artif. Intell. **97**(1–2), 245–271 (1997)
6. Chaboud, A.P., Chiquoine, B., Hjalmarsson, E., Vega, C.: Rise of the machines: algorithmic trading in the foreign exchange market. J. Finan. **69**(5), 2045–2084 (2015)
7. Chan, E.: Algorithmic Trading: Winning Strategies and Their Rationale, 1st edn. Wiley Publishing, Hoboken (2013)
8. Chen, T., Guestrin, C.: XGBoost: a scalable tree boosting system. CoRR abs/1603.02754 (2016)
9. Diakonikolas, J., Orecchia, L.: Alternating randomized block coordinate descent. In: Proceedings of the 35th International Conference on Machine Learning. Proceedings of Machine Learning Research, PMLR, 10–15 July 2018, Stockholmsmässan, Stockholm, Sweden (2018)
10. Fung, S.P.Y.: Optimal online two-way trading with bounded number of transactions. CoRR (2017)
11. Gaudel, R., Sebag, M.: Feature Selection as a one-player game. In: International Conference on Machine Learning, Haifa, Israel, pp. 359–366, June 2010. https://hal.inria.fr/inria-00484049
12. Di Graziano, G.(D.B.A.): Optimal trading stops and algorithmic trading. SSRN (2014). https://ssrn.com/abstract=2381830
13. Goldstein, M., Viljoen, T., Westerholm, P.J., Zheng, H.: Algorithmic trading, liquidity, and price discovery: an intraday analysis of the SPI 200 futures. Fin. Rev. **49**(2), 245–270 (2014)
14. Guyon, I., Elisseeff, A.: An introduction to variable and feature selection. J. Mach. Learn. Res. **3**, 1157–1182 (2003)
15. Hastie, T., Tibshirani, R., Friedman, J.: The Elements of Statistical Learning. SSS. Springer, New York (2009). https://doi.org/10.1007/978-0-387-84858-7
16. Kira, K., Rendell, L.A.: A practical approach to feature selection. In: Proceedings of the Ninth International Workshop on Machine Learning. ML92, San Francisco, CA, USA (1992)
17. Kirilenko, A., Kyle, A.S., Samadi, M., Tuzun, T.: The flash crash: high-frequency trading in an electronic market. J. Finan. **72**, 967–998 (2017)
18. Kohavi, R., John, G.H.: Wrappers for feature subset selection. Artif. Intell. **97**(1), 273–324 (1997)
19. Koller, D., Sahami, M.: Toward optimal feature selection. In: Proceedings of the Thirteenth International Conference on International Conference on Machine Learning, pp. 284–292. ICML 1996, San Francisco, CA, USA. Morgan Kaufmann Publishers Inc., Burlington (1996)

20. Labadie, M., Lehalle, C.A.: Optimal algorithmic trading and market microstructure. Working papers (2010)
21. Mangal, A., Holm, E.A.: A comparative study of feature selection methods for stress hotspot classification in materials. ArXiv e-prints, April 2018
22. Mitra, P., Murthy, C.A., Pal, S.K.: Unsupervised feature selection using feature similarity. IEEE Trans. Pattern Anal. Mach. Intell. **24**(3), 301–312 (2002)
23. Nesterov, Y.: Efficiency of coordinate descent methods on huge-scale optimization problems. SIAM J. Optim. **22**(2), 341–362 (2012)
24. Ng, A.Y.: Feature selection, l1 vs. l2 regularization, and rotational invariance. In: ICML 2004 (2004)
25. Staniak, M., Biecek, P.: Explanations of model predictions with live and breakdown packages. R J. **10**, 395 (2018)
26. Tibshirani, R.: Regression shrinkage and selection via the Lasso. J. Roy. Stat. Soc. B **58**, 267–288 (1994)
27. Tuv, E., Borisov, A., Runger, G., Torkkola, K.: Feature selection with ensembles, artificial variables, and redundancy elimination. J. Mach. Learn. Res. **10**, 1341–1366 (2009)
28. Vezeris, D., Kyrgos, T., Schinas, C.T.P., Loss, S.: Trading strategies comparison in combination with an MACD trading system. J. Risk Fin. Manage. **11**, 56 (2018)
29. Wright: Coordinate descent algorithms. Math. Program. **151**(1), 3–34 (2015)
30. Zarshenas, A., Suzuki, K.: Binary coordinate ascent: an efficient optimization technique for feature subset selection for machine learning. Knowl.-Based Syst. **110**, 191–201 (2016)
31. Zhang, T.: Adaptive forward-backward greedy algorithm for sparse learning with linear models. Curran Associates, Inc. (2009)

How Much Does Stock Prediction Improve with Sentiment Analysis?

Frederico G. Monteiro and Diogo R. Ferreira[✉]

Instituto Superior Técnico (IST), Universidade de Lisboa, 1049-001 Lisboa, Portugal
{frederico.monteiro,diogo.ferreira}@tecnico.ulisboa.pt

Abstract. Financial markets, such as the stock exchange, are known to be extremely volatile and sensitive to news published in the media. Using sentiment analysis, as opposed to using time series alone, should provide a better indication for the prospects of a given financial asset. In this work, the main goal is to quantify the benefit that can be obtained by adding sentiment analysis to predict the up or down movement of stock returns. The approach makes use of several different deep learning models, from vanilla models that rely on market indicators only, to recurrent networks that incorporate news sentiment as well. Surprisingly, the results suggest that the added benefit of sentiment analysis is diminute, and a more significant improvement can be obtained by using sophisticated models with advanced learning mechanisms such as attention.

Keywords: Stock prediction · Sentiment analysis · Deep learning · Attention

1 Introduction

Financial markets are part of the core structure of modern society, since the stability and prosperity of these markets have a great impact on economic development. Stock prediction, i.e. the prediction of the future value of a company stock, is an extremely difficult task because the extent to which the future price of a stock can be predicted from its past history has always been the source of much debate [1].

In essence, there are three major theories about stock behavior: the Random Walk Theory [1] endorsing that markets are completely arbitrary and stock prices are not predictable; the Elliott Wave Theory (EWT) [2] defending that trends follow repeating patterns and those patterns can be used to predict future movements; and the Efficient Market Hypothesis (EMH) [3] advocating that stock prices are by themselves the result of all available information and it is impossible to derive sustained gains.

Financial data plays a key role in all of the above theories, and its analysis is divided in two categories. Technical Analysis (TA), supporting EWT, is frequently employed in short-term approaches, as it tries to find the correct timing to enter and exit the market; however, it is somewhat subjective in nature, since it depends on the interpretation of technical charts and data. The other category

© Springer Nature Switzerland AG 2021
V. Bitetta et al. (Eds.): MIDAS 2020, LNAI 12591, pp. 16–31, 2021.
https://doi.org/10.1007/978-3-030-66981-2_2

is Fundamental Analysis (FA), supporting EMH, an approach that focuses more on the long-term by analyzing the intrinsic value of information, but is also hard to formalize into rules, with the challenge being how to handle unstructured data in a systematic manner.

It is interesting to observe how the two approaches above can be mapped to two fundamentally different data sources. On one hand, there are market indicators that reflect the day-to-day evolution of stock prices and can be used for TA. On the other hand, there are media news and sentiments that ultimately contribute to shape the evolution of those stocks, and are important for FA. While TA can be addressed, for example, with Time Series Prediction (TSP), FA may require the use of additional data sources and techniques, e.g. Sentiment Analysis (SA), to fully understand how the current stock price synthesizes all the information available about a company.

But how much can one gain by incorporating SA over a pure technical analysis such as TSP? In principle, SA should be able to provide some explanation for the seemingly random evolution of a time series, so it should be possible to achieve an increase in the accuracy of any stock prediction model. The question that we address in this paper is how much of an improvement can one actually obtain by incorporating SA. Since recently there has been an increasing focus on addressing both TSP and SA with Deep Learning (DL) models, we also use DL to facilitate the development of prediction models that combine both components.

In particular, and contrary to our initial intuition, we find that the choice of model and learning mechnisms, namely recurrence and attention, play a far greater role in improving the prediction accuracy than the inclusion of SA, which provides only a minor improvement. Although this might be somewhat unexpected, it does provide some confidence that the continued development and increasing sophistication of DL models will be able to extract useful predictions from market and news data.

In the following, Sect. 2 provides an overview of related work, Sect. 3 presents and discusses our implemented models and results, and Sect. 4 concludes the paper.

2 Related Work

Numerous attempts have been made to predict the movement of stock prices. Since stocks are in a continuous shaping process due to new information [4], their associated data is usually complex, uncertain, incomplete, and vague [5], with neither stationary (mean, variance, and frequency change over time) nor linear properties [6]. Stock prediction is acknowledged to be one of the most challenging time series tasks [7], with the hardest part being to select relevant features and learning mechanisms.

In this section, we have a look at how other authors have approached stock prediction from two different perspectives, namely from the perspective of TSP and from the perspective of SA, which are the most relevant components for our purpose. We also briefly discuss how Deep Learning can be used to address both components.

2.1 Time Series Prediction

Stock prediction has been often approached with statistical techniques, using models such as AR, MA, ARMA and ARIMA [8]. Typically, these models make predictions about a single stock and generally assume that it has linear properties. The main point here is that these models lack the capability to perceive non-linear behavior and dynamics between stocks [9].

Machine Learning (ML) has brought substantial advances in forecasting multiple stocks and understanding the hidden relationships between them. The most commonly used methods used are Support Vector Machines (SVM) [10], Random Forests [11], Bayesian Networks [12] and, more recently, Deep Learning (DL) [13].

The frequent use of DL and Neural Networks (NN) for stock prediction is justified by their ability to generalize the prediction to other assets and the relative ease when dealing with complex and noisy data. These models outperform statistical techniques [14,15] and other ML models as well [11,16].

Another alternative is Reinforcement Learning (RL), where an agent automates trades and reduces costs by systematizing TA as a set of rules to anticipate future price shifts [17]. This has been shown to achieve satisfactory results [18]. However, due to high volatility and noise, there might be significant differences between the patterns observed during training and testing, causing the agent to under-perform.

Table 1 presents a summary of the methods employed, the stock indexes and the data sources used by different authors. This survey leads us to conclude that *Yahoo Finance* and *Thomson Reuters* have been the most popular data sources; other sources used are *Bloomberg*, *Google Finance* and public datasets. On the other hand, the most common indexes and markets are the *American* (*S&P*, *NASDAQ*, *DJIA* and *NYSE*) and the *Asian* (*NSE*, *Nikkei*, *KOSPI* and *CSI*); others include *EuroStoxx50* and *Ibovespa*.

Table 1. Techniques and data sources for stock prediction based on time series.

Technique	Index	Yahoo finance	Thomson reuters	Other	Unidentified
Stats	American	[19–22]	[23,24]		[25]
	Asian		[24,26]		[13,27–29]
ML	American	[12,22,30–34]	[11,23,24]	[16,35,36]	[25,37,38]
	Asian	[12,39]	[24,26]	[40]	[28,38,41–43]
	Other	[12]		[44]	
NN	American	[22,31,32,34,45–47]	[11,23,48,49]	[16,50,51]	[25,37,38]
	Asian	[46]	[49]	[40]	[13,27–29,38,41,42,52]
RL	American	[18,53,54]	[54]		
	Asian	[54]	[54]		
	Other	[54]	[54]		

Regarding the time frame for training and evaluation, the shortest period found in these works was one month [30] and the longest was 66 years [37]. It is worth noting that statistical techniques tend to use shorter time frames to avoid noise and dimensionality issues [19–22], while ML techniques tend to use longer time frames to properly train the models [11,23,37,49]. Also, the directional movement of stocks (up or down) is the most common prediction goal, and arguably the most important one [35]. Only two articles tried to predict the actual stock price [36,47].

2.2 Sentiment Analysis

While SA can be regarded as a subfield of Natural Language Processing (NLP), its fast growth in recent years [55] has been propelled by numerous applications that transform human-generated information (news, tweets, etc.) about any topic (companies, products, politics, etc.) into a sentiment signal (usually positive or negative).

In stock markets, participants take actions (to buy, hold or sell stocks) that are defined and affected by what they read and by what those surrounding them read and share, including the opinions of sources they trust, which are also influenced by market news. From this point of view, the emotional sentiment upon a particular stock or company has become a fundamental part of stock prediction [56,57], and many authors have included this additional component in their prediction models.

Table 2 provides an overview of the related works using SA. The data sources are quite diversified, with the most popular being *News* and *Twitter*, which can be explained by the fact that they provide a simple way to access and collect data. In addition, *News* have the advantage of being created by accredited authors and

Table 2. Techniques and data sources for stock prediction based on sentiment analysis.

Model	Source					
	News	Social media	Twitter	Blog & Forums	Financial reports	None
Stats	[30,42,60]	[26,61]	[19,20,22]			[13,25,27,62–67], [14,15,24,28,29]
KNN	[60,68]	[68]		[68]		[66]
Naïve Bayes	[68]	[68]	[69]	[68]	[58]	[12]
MLP	[42,70]		[70]			[25,37,38,41]
Random forest	[34]	[71]	[35,69]		[58]	[16,23,24,43,66]
SVM	[31,36,39,72], [32,34,68,73]	[26,68,71]	[22,69,72]	[40,68]	[33]	[10,16,66,67,74,75], [24,28,40,44,76,77]
NN	[31,34,60]	[71]	[22,47]			[13,16,23,49,52,63, 78,79], [25,27,64–67,74,75], [11,14,15,28,46,48]
CNN	[32,45,50,70,80–82]		[70,80,81]			[28,29,37,38,41,65]
RNN	[32,42]					[28,29,37,38,49]
LSTM	[42,50,51,72,80,83]	[61]	[72,80]	[40]		[23,27– 29,38,40,41,49,84]
RL						[18,53,54,85,86]

newspapers [58], while *Twitter* can be used to more easily perceive sentiment, even though tweets are not suited for longer pieces or fully justified opinions on a topic [59].

Regarding sentiment polarity, in its simplest form it involves just two classes: positive and negative. A simple extension is ternary polarity by adding a neutral class [87]. More fined-grained SA can be achieved by categorizing flavors of feelings (such as *fear, joy, interest*) and emotions (such as *pleasantness, attention, sensitivity, aptitude*) [88], as well as their level of intensity [56]. Some authors [89] have also used a continuous value in a certain range, e.g. $[-1, +1]$.

An interesting conclusion from our survey is that private information causes small or insignificant changes in stock price [90]; in contrast, public information can cause major changes [91]. Therefore, it can be concluded that prediction accuracy is primarily influenced by public information. However, with private information it is still possible to improve the accuracy a bit further.

Another important conclusion is that markets tend to overreact when presented with negative news [92], and it is easier to predict downward trends than upward trends [93]. In general, pessimism tends to increase the trading volume and lead to predictions of negative returns [94], but it tends to disappear within one week.

Since each work in Table 2 uses a different dataset, the prediction models cannot be compared directly, even if their evaluation results are available. However, it is fair to say that prediction accuracy tends to increase from the top row to the bottom rows. In general, it is not easy to integrate SA features into statistical models, KNN, Naïve Bayes, MLP, Random Forests and SVM due to sparsity issues. Hence, they achieve lower accuracy than DL models. On the other hand, a Long Short-Term Memory (LSTM) incorporating SA registered the highest directional accuracy [40]. For us, this is not surprising since RNNs are especially appropriate for NLP and SA tasks.

2.3   Deep Learning

In recent years, as computer hardware allowed for training larger, deeper and more sophisticated neural networks, Deep Learning has become the state-of-the art approach in many fields, including stock prediction.

A Neural Network (NN) is composed of nodes and activation functions. The simplest architecture is the Feed-Forward Neural Network (FNN), which is based on the Multi-Layer Perceptron (MLP), and its name derives from the fact that information flows one-way, from input to output, across a set of densely-connected layers. A different model is the Convolutional Neural Network (CNN), where each layer performs a computation in the form of a sliding window over its input. CNNs are very popular in image processing where the input is 2D, but they are also very successful at processing 1D input such as text sentences [95] and time series [96].

For sequence processing, Recurrent Neural Networks (RNNs) [97] are very appropriate, since they keep an internal state and have a feedback loop to use that internal state as an additional input at each time step. For stock prediction, this means that it is possible to reuse information from the past to predict the future. However, simple RNNs are unable to perform well when a long-term context is required [98]. The LSTM architecture [99] improves on long-term dependencies by keeping a cell state with the possibility of carrying it across multiple time steps. More recently, slightly different versions have been proposed, such as the Gated Recurrent Unit (GRU) [100]. However, the LSTM can still provide state-of-the-art results [41].

Although significant improvements have been obtained with LSTM models, there are still limitations such as the vanishing gradient problem [101]. The use of attention mechanisms mitigates this constraint [102] by selectively retrieving the most relevant information from hidden states [103]. The underlying rationale is to increase the focus on important parts of the input instead of just trying to remember it afterwards [104]. Results suggest that the use of attention can capture long-term dependencies and outperform stand-alone LTSMs [51].

One way to use attention is to attach this mechanism to an LSTM or, alternatively, to develop a model similar to the Transformer [105] which relies exclusively on attention, without recurrence. However, for sequential data it is useful to keep track of the order between time steps, and the Transformer achieves this by means of positional encoding. In our context, the Transformer architecture can be simplified: for example, for time series data, the embedding layers that are typically used for NLP tasks can be removed; also, the decoder part of the Transformer can be removed to keep only the encoder as a prediction model. Another possibility is to use Multi-Head Attention, which essentially consists in combining multiple layers of attention in parallel to enhance the extraction of long-dependencies from the input sequence.

3   Stock Prediction with Sentiment Analysis

In this section, we report on a series of experiments that we performed during the Kaggle competition *Two Sigma: Using News to Predict Stock Movements.*[1] One of the key aspects of this competition is that it provided two different data sources: (1) market data, with typical indicators about stock prices, trading volumes and calculated returns; and (2) news data, with pre-calculated sentiment, novelty and volume counts for each news item. This made it possible to test several different models for stock prediction, both with and without sentiment analysis.

3.1   Data Description

The original dataset contained market and news data from the beginning of 2007 to the end of 2016. The market data included information on ~3000 US-listed companies, containing over 4 million samples with features such as date, asset code, asset name, daily open and close prices, daily trading volume, and open-to-open and close-to-close returns. These returns were calculated both daily and

---

[1] https://www.kaggle.com/c/two-sigma-financial-news.

for a 10-day period, and also both in raw and market-residualized form (i.e. by removing the movement of the market as a whole and leaving only the movement inherent to the asset).

On the other hand, the news data contained about 9 million samples (daily news items) regarding the same companies or related ones. For each news item, the sentiment was given as a probability distribution over 3 classes: positive, neutral and negative. Measures for the novelty and volume counts of each news item were also available, where novelty was calculated by comparing the asset-specific content of a news item against a cache of previous news items, and the volume was calculated by counting how many news items mentioned the asset within a certain time frame. However, for this work we used only the sentiment features of each news item.

As for the target variable, the competition used the open-to-open market-residualized return over a 10-day period into the future. In our experiments, and following common practice in the literature, we used only the directional movement (up or down) of this target variable. The problem then becomes a binary classification, which makes it especially easy to evaluate accuracy as the percentage of correct predictions, while also enabling the use of a standard loss function for training, i.e. binary cross-entropy.

3.2   Data Preprocessing

When joining the market data with the news data, it may be the case that there are several news items for a given asset on a given date. In this case, we group those news items together and compute their average sentiment.

On the other hand, it may be the case that there are actually no news items for a given asset on a particular date. In this case, we have tried three different approaches to fill in the sentiment for the asset, namely:

- *propagation*, where we use the sentiment of the previous date as the current sentiment, hence propagating the sentiment across multiple days;
- *balancing*, where we use a uniform probability distribution over the three classes (positive, neutral and negative, with $\frac{1}{3}$ probability for each);
- *neutralizing*, where we assign zero probability to the positive and negative classes, and 1.0 to the neutral class.

After joining both data sources according to one of the strategies above, the dataset was split into 2007–2014 for training, 2015 for validation, and 2016 for testing.

3.3   Prediction Models

In our models, we used 2 categorical features and 15 numerical features as input. The 2 categorical features were asset code and asset name, because each asset name (i.e. company) may have several asset codes, but each asset code belongs to a single company. As for the numerical features, 12 of those features come from the market data, and the remaining ones correspond to the 3 sentiment classes from news data.

The way these categorical and numerical features have been combined is illustrated in Fig. 1. In essence, the categorical features go through an embedding layer before being concatenated together with the numerical features. This initial block provides the input for all subsequent models, except for the Feed-Forward Neural Network (FNN) which uses a single time-step rather than 10 time-steps as the other models.

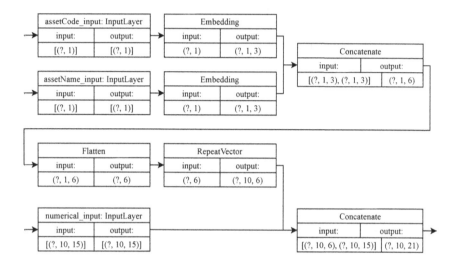

Fig. 1. Input block.

The first experiment, with a simple FNN (Fig. 2), already provides an interesting conclusion: when using this model, the sentiment features do not have any influence in the output predictions. In our view, this is due to the fact that the FNN does not incorporate any information on past behavior. Both with and without the sentiment features, this model scored a test accuracy of 53.3%.

Using a Convolutional Neural Network (CNN) (Fig. 3) with a 10 time-step window into the past improved the prediction accuracy. However, the test accuracy was higher without the sentiment features (55.1%) than with those features included (54.8%). We attribute this fact to overfitting, and dropout did not help in this case.

As soon as we moved on to recurrent networks (Fig. 4), we began to observe more consistent results. A simple Recurrent Neural Network (RNN) provided the same accuracy (55.2%) with and without sentiment features, but a Long Short-Term Memory (LSTM) improved from 55.2% to 55.3% when the sentiment features were included. We conclude that recurrence mechanisms are essential in order to leverage information from the past. However, a bi-directional LSTM did not improve the results, which is understandable, since the most recent history is probably more relevant for prediction than the distant past, so only the forward direction is useful.

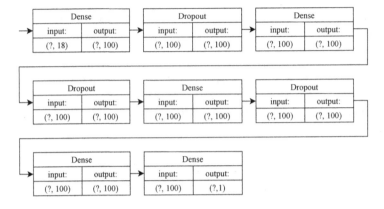

Fig. 2. FNN model.

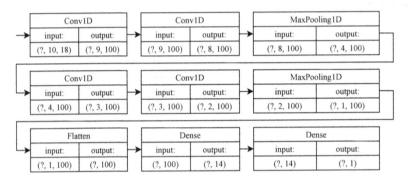

Fig. 3. CNN model.

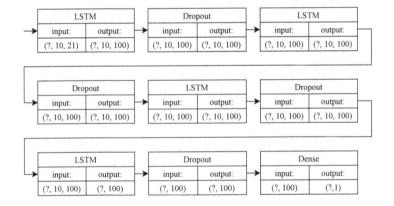

Fig. 4. LSTM model.

The accuracy improves even further when attention mechanisms are employed. An LSTM model with additive attention achieves 59.8% accuracy without sentiment features, and reaches 60.0% when those features are included. Moreover, a bidirectional LSTM with attention (Fig. 5) improves that result even further to 60.4%. Here, bidirectionality is definitely beneficial because it helps the attention mechanism learn which time-steps are the most relevant for prediction.

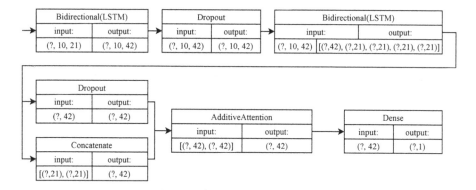

Fig. 5. Bidirectional LSTM with attention, the best performing model.

We have also experimented with a Transformer-like architecture, which relies on attention but not recurrence. Here the results were inferior, with 55.6% accuracy without sentiment features, and an increase to 55.7% when those features were included.

Table 3 provides an overview of the results, where it becomes apparent that the inclusion of sentiment features can account for an improvement of at most 0.6% in test accuracy, while the use of recurrence and attention mechanisms can provide an improvement of 5.6% over a baseline CNN that uses the same input data. Other metrics, such as area under the ROC curve (AUC) and Matthews correlation coefficient (MCC) have commensurate improvements with accuracy.

Table 3. Comparison of accuracy.

Model	Market only	Market+News propagated	Market+News balanced	Market+News neutralized
FNN	**0.533**	0.532	0.533	0.533
CNN	**0.551**	0.543	0.543	0.548
RNN	0.552	0.547	**0.552**	0.550
LSTM	0.552	0.551	**0.553**	0.551
Bi-LSTM	0.543	0.548	0.547	**0.549**
LSTM + Att	0.598	0.599	**0.600**	0.600
Bi-LSTM + Att	0.600	0.603	0.602	**0.604**
Transformer	0.556	0.552	**0.557**	0.550

All models have been implemented with TensorFlow and have been tested with different hyper-parameters (number of layers, number of filters, kernel sizes, dropout rates, etc.). Here we reported the results achieved with the best-performing version.

4   Conclusion

From the literature review that we presented in Sect. 2, we expected that sentiment analysis would improve stock prediction, but our experiments and results in Sect. 3 suggest that the improvement is rather modest compared to our initial expectation.

On the other hand, our findings point to a possible avenue for the continued improvement of prediction models and their accuracy. By making use of learning mechanisms such as recurrence and attention, and others that may appear along the way, it is possible to keep improving the results. It is also apparent that such mechanisms will work better in combination rather than in isolation; specifically, recurrence with attention works better than either of those mechanisms alone.

In addition, we found that propagating sentiment by keeping past sentiment when there are no news is not the best approach. In this respect, the market seems to forget past sentiment in a matter of days, to the point that it is no longer possible to clearly determine the current sentiment as being positive, negative, or neutral.

In this work, our prediction target was binary, in the form of a directional movement. In future work, we plan to apply similar models for regression tasks such as predicting price and volatility. In addition, the kind of evaluation that we provide here is by no means the end of the story; to derive actual benefits in the real-world, other components, such as a trading strategy to enter and exit the market, are necessary.

References

1. Fama, E.F.: The behavior of stock-market prices. J. Bus. **38**(1), 34–105 (1965)
2. Elliott, R.N.: The Wave Principle. Alanpuri Trading, Rancho Cucamonga (1938)
3. Fama, E.F.: Efficient capital markets: a review of theory and empirical work. J. Finan. **25**(2), 383–417 (1970)
4. LeBaron, B., Arthur, W.B., Palmer, R.: Time series properties of an artificial stock market. JEDC **23**(9–10), 1487–1516 (1999)
5. Emerson, S., Kennedy, R., O'Shea, L., O'Brien, J.: Trends and applications of machine learning in quantitative finance. In: ICEFR 2019, June 2019
6. Vanstone, B., Finnie, G.: An empirical methodology for developing stock market trading systems using artificial neural networks. Expert Syst. Appl. **36**(3), 6668–6680 (2009)
7. Hussain, A.J., Knowles, A., Lisboa, P.J.G., El-Deredy, W.: Financial time series prediction using polynomial pipelined neural networks. Expert Syst. Appl. **35**(3), 1186–1199 (2008)

8. Box, G.E.P., Jenkins, G.: Time Series Analysis, Forecasting and Control, 5th edn. Wiley, Hoboken (1990)
9. Fang, J., Jacobsen, B., Qin, Y.: Predictability of the simple technical trading rules: an out-of-sample test. Rev. Fin. Econ. **23**(1), 30–45 (2014)
10. Cao, L.J., Tay, F.E.H.: Support vector machine with adaptive parameters in financial time series forecasting. IEEE Trans. Neural Netw. **14**(6), 1506–1518 (2003)
11. Krauss, C., Do, X.A., Huck, N.: Deep neural networks, gradient-boosted trees, random forests: statistical arbitrage on the S&P 500. EJOR **259**(2), 689–702 (2017)
12. Malagrino, L.S., Roman, N.T., Monteiro, A.M.: Forecasting stock market index daily direction: a Bayesian network approach. Expert Syst. Appl. **105**, 11–22 (2018)
13. Chong, E., Han, C., Park, F.C.: Deep learning networks for stock market analysis and prediction: methodology, data representations, and case studies. Expert Syst. Appl. **83**, 187–205 (2017)
14. Kohzadi, N., Boyd, M.S., Kermanshahi, B., Kaastra, I.: A comparison of artificial neural network and time series models for forecasting commodity prices. Neurocomputing **10**(2), 169–181 (1996)
15. Zhang, G.P.: Time series forecasting using a hybrid ARIMA and neural network model. Neurocomputing **50**, 159–175 (2003)
16. Chatzis, S.P., Siakoulis, V., Petropoulos, A., Stavroulakis, E., Vlachogiannakis, N.: Forecasting stock market crisis events using deep and statistical machine learning techniques. Expert Syst. Appl. **112**, 353–371 (2018)
17. Nazário, R.T.F., e Silva, J.L., Sobreiro, V.A., Kimura, H.: A literature review of technical analysis on stock markets. QREF **66**, 115–126 (2017)
18. Deng, Y., Bao, F., Kong, Y., Ren, Z., Dai, Q.: Deep direct reinforcement learning for financial signal representation and trading. IEEE Trans. Neural Netw. Learn. Syst. **28**(3), 653–664 (2017)
19. Si, J., Mukherjee, A., Liu, B., Li, Q., Li, H., Deng, X.: Exploiting topic based twitter sentiment for stock prediction. In: ACL 2013, vol. 2, pp. 24–29, August 2013
20. Si, J., Mukherjee, A., Liu, B., Pan, S.J., Li, Q., Li, H.: Exploiting social relations and sentiment for stock prediction. In: EMNLP, pp. 1139–1145, October 2014
21. Dickinson, B., Hu, W.: Sentiment analysis of investor opinions on Twitter. Soc. Netw. **4**(3), 62–71 (2015)
22. Mittal, A., Goel, A.: Stock prediction using twitter sentiment analysis. Standford University, CS229 (2012)
23. Fischer, T., Krauss, C.: Deep learning with long short-term memory networks for financial market predictions. EJOR **270**(2), 654–669 (2018)
24. Lee, T.K., Cho, J.H., Kwon, D.S., Sohn, S.Y.: Global stock market investment strategies based on financial network indicators using machine learning techniques. Expert Syst. Appl. **117**, 228–242 (2019)
25. Guresen, E., Kayakutlu, G., Daim, T.U.: Using artificial neural network models in stock market index prediction. Expert Syst. Appl. **38**(8), 10389–10397 (2011)
26. Wang, H., Lu, S., Zhao, J.: Aggregating multiple types of complex data in stock market prediction: a model-independent framework. Knowl.-Based Syst. **164**, 193–204 (2019)
27. Kim, H.Y., Won, C.H.: Forecasting the volatility of stock price index: a hybrid model integrating LSTM with multiple GARCH-type models. Expert Syst. Appl. **103**, 25–37 (2018)

28. Long, W., Lu, Z., Cui, L.: Deep learning-based feature engineering for stock price movement prediction. Knowl.-Based Syst. **164**, 163–173 (2019)
29. Selvin, S., Vinayakumar, R., Gopalakrishnan, E.A., Menon, V.K., Soman, K.P.: Stock price prediction using LSTM. In: ICACCI (RNN and CNN-sliding window model), September 2017
30. Schumaker, R.P., Chen, H.: Textual analysis of stock market prediction using financial news articles. In: AMCIS 2006, vol. 3, pp. 1422–1430, December 2006
31. Ding, X., Zhang, Y., Liu, T., Duan, J.: Using structured events to predict stock price movement: an empirical investigation. EMNLP **2014**, 1415–1425 (2014)
32. Vargas, M.R., De Lima, B.S.L.P., Evsukoff, A.G.: Deep learning for stock market prediction from financial news articles. CIVEMSA **2017**, 60–65 (2017)
33. Nguyen, T.H., Shirai, K.: Topic modeling based sentiment analysis on social media for stock market prediction. In: ACL-IJCNLP 2015, vol. 1, pp. 1354–1364, July 2015
34. Weng, B., Lu, L., Wang, X., Megahed, F.M., Martinez, W.: Predicting short-term stock prices using ensemble methods and online data sources. Expert Syst. Appl. **112**, 258–273 (2018)
35. Nasseri, A.A., Tucker, A., de Cesare, S.: Quantifying StockTwits semantic terms' trading behavior in financial markets: an effective application of decision tree algorithms. Expert Syst. Appl. **42**(23), 9192–9210 (2015)
36. Deng, S., Mitsubuchi, T., Shioda, K., Shimada, T., Sakurai, A.: Combining technical analysis with sentiment analysis for stock price prediction. In: DASC2011, pp. 800–807, December 2011
37. Di Persio, L., Honchar, O.: Artificial neural networks architectures for stock price prediction: comparisons and applications. Int. J. Circuits, Syst. Sig. Process. **10**, 403–413 (2016)
38. Hiransha, M., Gopalakrishnan, E.A., Menon, V.K., Soman, K.P.: NSE stock market prediction using deep-learning models. Procedia Comput. Sci. **132**, 1351–1362 (2018)
39. Li, X., Xie, H., Chen, L., Wang, J., Deng, X.: News impact on stock price return via sentiment analysis. Knowl.-Based Syst. **69**, 14–23 (2014)
40. Li, J., Bu, H., Wu, J.: Sentiment-aware stock market prediction: a deep learning method. In: ICSSSM 2017, June 2017
41. Kim, S., Kang, M.: Financial series prediction using attention LSTM. arXiv:1902.10877, February 2019
42. Akita, R., Yoshihara, A., Matsubara, T., Uehara, K.: Deep learning for stock prediction using numerical and textual information. In: ICIS 2016, vol. 1, pp. 1–6, June 2016
43. Zhang, J., Cui, S., Xu, Y., Li, Q., Li, T.: A novel data-driven stock price trend prediction system. Expert Syst. Appl. **97**, 60–69 (2017)
44. Paiva, F.D., Cardoso, R.T.N., Hanaoka, G.P., Duarte, W.M.: Decision-making for financial trading: a fusion approach of machine learning and portfolio selection. Expert Syst. Appl. **115**, 635–655 (2019)
45. Ding, X., Zhang, Y., Liu, T., Duan, J.: Deep learning for event-driven stock prediction. IJCAI **2015**, 2327–2333 (2015)
46. Zhou, F., Zhou, H.M., Yang, Z., Yang, L.: EMD2FNN: a strategy combining empirical mode decomposition and factorization machine based neural network for stock market trend prediction. Expert Syst. Appl. **115**, 136–151 (2019)
47. Bollen, J., Mao, H., Zeng, X.: Twitter mood predicts the stock market. J. Comput. Sci. **2**(1), 1–8 (2011)

48. Moews, B., Herrmann, J.M., Ibikunle, G.: Lagged correlation-based deep learning for directional trend change prediction in financial time series. Expert Syst. Appl. **120**, 197–206 (2019)
49. Baek, Y., Kim, H.Y.: ModAugNet: a new forecasting framework for stock market index value with an overfitting prevention LSTM module and a prediction LSTM module. Expert Syst. Appl. **113**, 457–480 (2018)
50. Prosky, J., Song, X., Tan, A., Zhao, M.: Sentiment predictability for stocks. arXiv:1712.05785, December 2017
51. Hollis, T., Viscardi, A., Yi, S.E.: A comparison of LSTMs and attention mechanisms for forecasting financial time series. arXiv:1812.07699, December 2018
52. Day, M.Y., Lee, C.C.: Deep learning for financial sentiment analysis on finance news providers. In: ASONAM, pp. 1127–1134, August 2016
53. Almahdi, S., Yang, S.Y.: An adaptive portfolio trading system: a risk-return portfolio optimization using recurrent reinforcement learning with expected maximum drawdown. Expert Syst. Appl. **87**, 267–279 (2017)
54. Jeong, G., Kim, H.Y.: Improving financial trading decisions using deep Q-learning: predicting the number of shares, action strategies, and transfer learning. Expert Syst. Appl. **117**, 125–138 (2019)
55. Mäntylä, M., Graziotin, D., Kuutila, M.: The evolution of sentiment analysis - a review of research topics, venues, and top cited papers. Comput. Sci. Rev. **27**, 16–32 (2018)
56. Devitt, A., Ahmad, K.: Sentiment polarity identification in financial news: a Cohesion-based approach. ACL **2007**, 984–991 (2007)
57. Feuerriegel, S., Prendinger, H.: News-based trading strategies. Decis. Support Syst. **90**, 65–74 (2016)
58. Sehgal, V., Song, C.: SOPS: stock prediction using web sentiment. ICDM **2007**, 21–26 (2007)
59. Cambria, E., Schuller, B., Xia, Y., Havasi, C.: New avenues in opinion mining and sentiment analysis. IEEE Intell. Syst. **28**(2), 15–21 (2013)
60. Wuthrich, B., Cho, V., Leung, S., Permunetilleke, D., Sankaran, K., Zhang, J.: Daily stock market forecast from textual web data. In: SMC 1998, vol. 3, pp. 2720–2725, October 1998
61. Xiong, R., Nichols, E.P., Shen, Y.: Deep learning stock volatility with google domestic trends. arXiv:1512.04916, December 2015
62. Agarwal, S., Kumar, S., Goel, U.: Stock market response to information diffusion through internet sources: a literature review. IJIM **45**, 118–131 (2019)
63. Arévalo, R., García, J., Guijarro, F., Peris, A.: A dynamic trading rule based on filtered flag pattern recognition for stock market price forecasting. Expert Syst. Appl. **81**, 177–192 (2017)
64. Hollis, T.: deep learning algorithms applied to blockchain-based financial time series. Technical report, University of Manchester (2018)
65. Hoseinzade, E., Haratizadeh, S.: CNNpred: CNN-based stock market prediction using a diverse set of variables. Expert Syst. Appl. **129**, 273–285 (2019)
66. Huang, C.J., Yang, D.X., Chuang, Y.T.: Application of wrapper approach and composite classifier to the stock trend prediction. Expert Syst. Appl. **34**(4), 2870–2878 (2008)
67. Huang, W., Nakamori, Y., Wang, S.Y.: Forecasting stock market movement direction with support vector machine. Comput. Oper. Res. **32**(10), 2513–2522 (2005)

68. Nassirtoussi, A.K., Aghabozorgi, S., Wah, T.Y., Ngo, D.C.L.: Text mining of news-headlines for FOREX market prediction: a multi-layer dimension reduction algorithm with semantics and sentiment. Expert Syst. Appl. **42**(1), 306–324 (2015)
69. Li, B., Chan, K.C.C., Ou, C., Ruifeng, S.: Discovering public sentiment in social media for predicting stock movement of publicly listed companies. Inf. Syst. **69**, 81–92 (2017)
70. Kar, S., Maharjan, S., Solorio, T.: RiTUAL-UH at SemEval-2017 Task 5: sentiment analysis on financial data using neural networks. In: SemEval-2017, pp. 877–882, August 2017
71. Weng, B., Ahmed, M.A., Megahed, F.M.: Stock market one-day ahead movement prediction using disparate data sources. Expert Syst. Appl. **79**, 153–163 (2017)
72. Cabanski, T., Romberg, J., Conrad, S.: HHU at SemEval-2017 Task 5: fine-grained sentiment analysis on financial data using machine learning methods. In: SemEval-2017, pp. 832–836, August 2017
73. Schumaker, R.P., Chen, H.: A quantitative stock prediction system based on financial news. Inf. Process. Manage. **45**(5), 571–583 (2009)
74. Kara, Y., Boyacioglu, M.A., Baykan, Ö.K.: Predicting direction of stock price index movement using artificial neural networks and support vector machines: the sample of the Istanbul stock exchange. Expert Syst. Appl. **38**(5), 5311–5319 (2011)
75. Kim, K.: Financial time series forecasting using support vector machines. Neurocomputing **55**(1–2), 307–319 (2003)
76. Sen, J., Chaudhuri, T.D.: Decomposition of time series data of stock markets and its implications for prediction - an application for the Indian auto sector. In: ABRMP 2016, January 2016
77. Zhiqiang, G., Huaiqing, W., Quan, L.: Financial time series forecasting using LPP and SVM optimized by PSO. Soft. Comput. **17**(5), 805–818 (2013)
78. Arévalo, A., Niño, J., Hernández, G., Sandoval, J.: High-frequency trading strategy based on deep neural networks. In: Huang, D.-S., Han, K., Hussain, A. (eds.) ICIC 2016. LNCS (LNAI), vol. 9773, pp. 424–436. Springer, Cham (2016). https://doi.org/10.1007/978-3-319-42297-8_40
79. Dixon, M., Klabjan, D., Bang, J.H.: Classification-based financial markets prediction using deep neural networks. Algorithmic Finan. **6**(3–4), 67–77 (2017)
80. Ghosal, D., Bhatnagar, S., Akhtar, M.S., Ekbal, A., Bhattacharyya, P.: IITP at SemEval-2017 Task 5: an ensemble of deep learning and feature based models for financial sentiment analysis. In: SemEval-2017, pp. 899–903, August 2017
81. Pivovarova, L., Escoter, L., Klami, A., Yangarber, R.: HCS at SemEval-2017 Task 5: sentiment detection in business news using convolutional neural networks. In: SemEval-2017, pp. 842–846, August 2017
82. Mansar, Y., Gatti, L., Ferradans, S., Guerini, M., Staiano, J.: Fortia-FBK at SemEval-2017 Task 5: bullish or Bearish? Inferring sentiment towards brands from financial news headlines. arXiv:1704.00939, April 2017
83. Moore, A., Rayson, P.: Lancaster A at SemEval-2017 Task 5: evaluation metrics matter: predicting sentiment from financial news headlines. arXiv:1705.00571, May 2017
84. Nelson, D., Pereira, A., de Oliveira, R.: Stock market's price movement prediction with LSTM neural networks. IJCNN 2017, May 2017
85. Bertoluzzo, F., Corazza, M.: Testing different reinforcement learning configurations for financial trading: introduction and applications. Procedia Econo. Finan. **3**, 68–77 (2012)

86. Pendharkar, P.C., Cusatis, P.: Trading financial indices with reinforcement learning agents. Expert Syst. Appl. **103**, 1–13 (2018)
87. Rojas-Barahona, L.M.: Deep learning for sentiment analysis. Lang. Linguist. Compass **10**(12), 701–719 (2016)
88. Loia, V., Senatore, S.: A fuzzy-oriented sentic analysis to capture the human emotion in Web-based content. Knowl.-Based Syst. **58**, 75–85 (2014)
89. Barbaglia, L., Consoli, S., Manzan, S.: Monitoring the business cycle with fine-grained, aspect-based sentiment extraction from news. In: Bitetta, V., Bordino, I., Ferretti, A., Gullo, F., Pascolutti, S., Ponti, G. (eds.) MIDAS 2019. LNCS (LNAI), vol. 11985, pp. 101–106. Springer, Cham (2020). https://doi.org/10.1007/978-3-030-37720-5_8
90. Ahern, K.R., Sosyura, D.: Who writes the news? Corporate press releases during merger negotiations. J. Finan. **69**(1), 241–291 (2014)
91. Vega, C.: Stock price reaction to public and private information. J. Financ. Econ. **82**(1), 103–133 (2006)
92. Niederhoffer, V.: The analysis of world events and stock prices. J. Bus. **44**(2), 193–219 (1971)
93. Hu, Y., Liu, K., Zhang, X., Su, L., Ngai, E.W.T., Liu, M.: Application of evolutionary computation for rule discovery in stock algorithmic trading: a literature review. Appl. Soft Comput. **36**, 534–551 (2015)
94. Tetlock, P.C.: Giving content to investor sentiment: the role of media in the stock market. J. Finan. **62**(3), 1139–1168 (2007)
95. Kim, Y.: Convolutional neural networks for sentence classification. arXiv:1408.5882, October 2014
96. Fawaz, H.I., Forestier, G., Weber, J., Idoumghar, L., Muller, P.: Deep learning for time series classification: a review. Data Min. Knowl. Disc. **33**(4), 917–963 (2019)
97. Sutskever, I., Martens, J., Hinton, G.E.: Generating text with recurrent neural networks. In: ICML 2011, pp. 1017–1024 (2011)
98. Bengio, Y., Simard, P., Frasconi, P.: Learning long-term dependencies with gradient descent is difficult. IEEE Trans. Neural Netw. **5**(2), 157–166 (1994)
99. Hochreiter, S., Schmidhuber, J.: Long short term memory. Neural Comput. **9**(8), 1735–1780 (1997)
100. Cho, K., van Merrienboer, B., Gulcehre, C., Bougares, F., Schwenk, H., Bengio, Y.: Learning Phrase Representations using RNN encoder-decoder for statistical machine translation. arXiv:1406.1078, June 2014
101. Hochreiter, S.: The Vanishing gradient problem during learning recurrent neural nets and problem solutions. In: IJUFKS 1998 **6**(2), pp. 107–116 (1998)
102. Lin, Z., et al.: A structured self-attentive sentence embedding. In: ICLR 2017, March 2017
103. Bahdanau, D., Cho, K., Bengio, Y.: Neural machine translation by jointly learning to align and translate. arXiv:1409.0473 (2014)
104. Sutskever, I., Vinyals, O., Le, Q.V.: Sequence to sequence learning with neural networks. In: NIPS 2014, pp. 3104–3112 (2014)
105. Vaswani, A., et al.: Attention is all you need. In: NIPS 2017, pp. 6000–6010, December 2017

Applying Machine Learning to Predict Closing Prices in Stock Market: A Case Study

Matteo Greco[1], Michele Spagnoletta[2], Annalisa Appice[1(✉)],
and Donato Malerba[1]

[1] University of Bari Aldo Moro, 700125 Bari, Italy
m.greco104@studenti.uniba.it, {annalisa.appice,donato.malerba}@uniba.it
[2] Exprivia S.p.A, 70056 Molfetta, Italy
michele.spagnoletta2@exprivia.com

Abstract. The stock's closing price is the standard benchmark used by investors to track the stock performance over time. In particular, understanding the trend of the stock's closing prices is of fundamental importance to choose investments carefully. In this paper, we address the task of forecasting closing prices of Exprivia S.p.A.'s stock by comparing the performance of both traditional and deep machine learning methods. Preliminary experiments show that the multi-variate setting can significantly outperform the univariate one and that deep learning can gain accuracy compared to traditional machine learning methods in the considered task.

Keywords: Stock's closing price forecasting · Machine learning · Deep learning · Time series analysis

1 Introduction

The closing price for any stock is the final price at which it trades during regular market hours on any given day. This is a useful marker for investors to assess changes in stock prices over time [1]. In particular, the closing price on one day can be compared to the closing price on the previous days, in order to measure the changes in market sentiment towards that stock. In this scenario, the company announcements are typically released after the close of the regular trading day, in order to give traders a chance to digest the news before acting upon it. So, one pitfall of closing prices of a company is that they will not usually reflect any news released by the company that day. In addition, the release of news generally causes a stock's price to move dramatically up or down over consecutive days by adding complexity to task of predicting the future of stock's price. On the other hand, accurate forecasts of closing prices are very important to estimate the success or bankruptcy of a company and to realize the portfolio optimization.

Several studies have already investigated the application of statistical and machine learning techniques for stock market prediction (e.g. [3,5,7,11]).

© Springer Nature Switzerland AG 2021
V. Bitetta et al. (Eds.): MIDAS 2020, LNAI 12591, pp. 32–39, 2021.
https://doi.org/10.1007/978-3-030-66981-2_3

In this paper, we focus the attention on investigating the achievements of various machine learning methods in the task of forecasting the closing price of Exprivia S.p.A.'s stock.[1] by looking for ancillary variates that may actually aid in accuracy improvements. In fact, we formulate the forecasting problem in a time series setting and compare the performance of both an univariate and a multi-variate solution to the problem. In the univariate setting, the forecasting model is trained to forecast closing prices based on closing prices of the company measured on the previous days. In the multi-variate setting, various ancillary variates have been identified so that the forecasting model can be trained to forecast the closing price of Exprivia S.p.A.'s stock based on the closing price and these ancillary variates.

The main contribution of this preliminary study is the investigation of which possible ancillary covariates may be accounted for gaining accuracy and the comparative evaluation of the accuracy of traditional and deep machine learning methods in the considered scenario. In particular, we compare the performance achieved training ARIMA [2] in the univariate setting, as well as VAR [12], Random Forest [10], XGBoost [4] and LSTM [8] in the multi variate forecasting. The empirical study has been performed processing data collected between October 12, 2015 and November 10, 2019. Following the boom in deep learning, the results confirm that multi-variate setting coupled with deep learning architecture can yield more accurate forecasts than statistical and machine learning competitors in this task.

The paper is organized as follows. In the next Section, we describe the machine learning algorithms considered for the study. Section 3 is devoted to illustrate data, experimental set-up and results. Finally, Sect. 4 draws conclusions and sketches future directions of this work.

2   Forecasting Algorithms

In the following we assume that $\mathbf{Z}$ denotes a variable vector populated with both the target variate – closing price – Y and the ancillary variates $A_1, \ldots, A_n$. These variable have been daily measured for the same company yielding multi-variate time series. t denotes a time point.

In the univariate setting, the forecasting model is trained considering the time series algorithm ARIMA. It predicts $y(t)$ according to the univariate model:

$$(1 - \hat{L})^d y(t) = c + \sum_{i=1}^{p} \phi(i) L^i (1 - L)^d y(t - i) + \sum_{i=1}^{q} \sigma(i) L^i \epsilon(t - i), \quad (1)$$

where $1 - L$ is the differencing operator, L is the lag operator, d is the order of the differencing operation, c is a constant, $\phi(i)$ and $\sigma(i)$ are coefficients, $y(t - i)$ is the time series value at time $t - i$ and $\epsilon(t - i)$ is the residual error at $t - i$. A characteristic of ARIMA is that it introduces this differencing operation to

---

[1] https://www.exprivia.it/en/?cl=1.

transform a non-stationary time series into a stationary one. This can be very useful when analyzing real-life time series data as closing prices. In this study, the best parameters (p, d, q) of ARIMA are determined by using the stepwise algorithm introduced by Hyndman and Khandakar [9]. This algorithm conducts a search over possible models within the order constraints provided and selects the model minimizing the BIC criterion.

In the multivariate setting, the forecasting model is trained considering: the time series algorithm VAR, the regression algorithms Random Forest and XGBoost, and the deep learning architecture LSTM.

The algorithm VAR trains a multi-variate forecasting model by taking into account the interactions among the multiple univariate time series defined in the variable vector $\mathbf{Z}$, that is:

$$\hat{\mathbf{z}}(t) = \hat{y}(t), \hat{a_1}(t), \ldots, \hat{a_n}(t) = \mathbf{c} + \sum_{i=1}^{p} \mathbf{\Pi}_i \mathbf{z}(t - i), \qquad (2)$$

where $\mathbf{\Pi}_i$ is the $((n + 1) \times (n + 1))$ coefficient matrix, $\mathbf{z}(t - i)$ is the $(n + 1) \times 1$ vector of values observed for the target variable Y and the ancillary variables $A1, \ldots, A_n$ at time $t - i$ and $\mathbf{c}$ is the $(n + 1) \times 1$ vector of the white noise error (serially uncorrelated or independent) with time invariant covariance matrix. We note that Eq. 2 comprises $n + 1$ multi-variate forecasting functions – one forecasting function for each variable in $\mathbf{Z}$ – comprising the multi-variate fore-casting model to forecast $\hat{y}(t)$. The best p is selected by minimizing the BIC criterion. Differently from ARIMA, the algorithm VAR assumes the time series stationary so that it does not apply any differencing operator, while it accounts for associations among multiple variates.

The Random Forest and XGBoost are robust, popular algorithms used in regression and classification, which are designed in the ensemble learning paradigm by resorting to bagging and boosting, respectively. In this study, the Random Forest algorithm has been run with the number of trees set equal to 100, the Gini Index as heuristic and the number of variates sampled set equal to the squared root of the number of independent variables. XGBoost has been run with the max tree depth set equal to 500. Both algorithms have been used to train a regression model from the training set extracted from the multivariate time series by using the sliding windowing operator. In particular, the multi-variate training time series have been divided into sliding, consecutive windows with window length w ($w = 5$ in this study). For each window, a sample is generated. In the window, $Y(w)$ represents the target variable to predict, while $Y(1), \ldots Y(w - 1), A_1(1), \ldots A_1(w), \ldots, A_n(1), \ldots A_n(w)$ represent the indepen-dent variables to be processed as predictors of the regression model.

Finally, the LSTM is a special case of a Recurrent Neural Network, which is commonly used in the sequence analysis. The cell states have the ability to remove or add information to the cell, regulated by gates. In theory, the presence of cell states makes the LSTMs able to handle the long-term dependency. We have defined an architecture with one LSTM state-full layer having 25 neurons[2],

---

[2] Preliminary experiments have been performed to set the size of this layer.

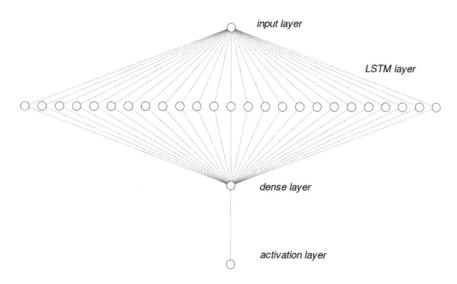

Fig. 1. LSTM architecture

Table 1. A sample of the considered data

Timestamp	Open	High	Low	Close	Adj Close	Volume
November 06, 2019	0.857	0.881	0.846	0.851	0.851	97204
November 07, 2019	0.851	0.885	0.829	0.839	0.839	60771
November 08, 2019	0.839	0.862	0.839	0.851	0.851	17553
November 09, 2019	0.842	0.863	0.842	0.859	0.859	84551
November 10, 2019	0.838	0.870	0.838	0.840	0.840	76150
November 11, 2019	0.836	0.866	0.836	0.866	0.866	27330
...	...	...	...	...	...	...

one dense layer and one activation layer with tangent activation function (Fig. 1). Data have been scaled between 0 and 1. In addition, we have set the batch size equal to 5, number of epochs equal to 400, the time step equal to 1 and the dropout equal to 0.00105. We have used the optimizer ADAM and the MSE as loss function.

3   Empirical Evaluation

In this section, we illustrate the results of our empirical validation of the performance of methods illustrated in Sect. 2. We start describing the data (Sect. 3.1). Then we introduce the experimental setting (Sect. 3.2). Finally, we illustrate the results (Sect. 3.3).

3.1   Data

We process the historical stock data of Exprivia S.P.A. daily collected by Yahoo Financial, between October 12, 2015 and October 11, 2019. This data collection comprises six timestamped, numeric variables. Every variable contributes to describe the stock market. Specifically,

- **Close** denotes the closing price as it is adjusted at the end of day;
- **Open** denotes opening price at the beginning of the day. Missing values have been assigned to the closing price at the previous day;
- **High** denotes the price achieved during the current day;
- **Low** denotes the minimum price achieved during the current day;
- **Volume** is the business volume on the day;
- **Adj Close** is a technical indicator that reveals the real value of stock measured by accounting for both stock split and dividends.

The variable **Close** is the target to forecast, while **Open, High, Low, Volume** and **Adj Close** are the ancillary variates of this study. **Close, Open, High, Low** and **Adj Close** assume values ranging between 0 and 2.15. **Volume** assumes values between 0 and 5172259. A sample of the considered data is reported in Table 1.

3.2   Experimental Setting

All algorithms have been implemented in Python 3.6 with libraries **Sklearn, pandas, statmodels, pmdarima, xgboost, tensorflow** and **plotly**.

We partition the data collection in training set (October 12, 2015-October 11, 2018) and testing set (October 12, 2018-October 11, 2019) To evaluate the ability of forecasting the closing price accurately, we measure the Root Mean Squared Error (RMSE) of forecasting models learned on the training set and used to predict the testing samples. Formally,

$$RMSE = \sqrt{\frac{\sum\limits_{t \in testing\ set} (y(\hat{t}) - y(t))^2}{|testing\ set|}}. \tag{3}$$

In the experiments with the algorithm VAR, we apply the Granger Causality Test [6], in order to select the ancillary variates which are casually relevant to the target one. So, we have identified **Open, High, Low** and **Adj Close**. As these variates are non-stationary they have been differentiated to make them stationary. In the experiments with the regression algorithms, no feature selection has been done.

3.3   Results

The RMSEs of ARIMA, VAR, Random forest, XGBoost and LSTM measured on the testing data are reported in Fig. 2. These results confirm that the multivariate setting outpefrom the univariate one. In fact, the univariate algorithm

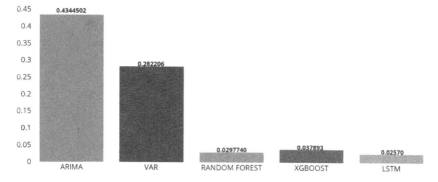

Fig. 2. RMSEs of ARIMA, VAR, Random Forest, XGBoost and LSTM computed in the testing period between October 12, 2018 and October 11, 2019.

ARIMA achieves the lowest performance (with RMSE = 0.4344) in this comparative study. In addition, the regression methods defined in traditional machine learning (Random forest and XGBoost) and deep learning (LSTM) outperform the time series approaches (ARIMA and VAR). The highest accuracy is achieved with the LSTM with RMSE equal to 0.0257. This result encourages us in continuing with deep learning in the scenario. In any case, the runner-up algorithm – Random Forest – also achieves competitive results (with RMSE = 0.0297).

The real closing prices and the forecasts yielded by LSTM and Random Forest respectively are plot in Fig. 3 for the testing period. Both algorithms yield forecasts that are very close to real data. Figure 4 shows the squared residuals of both LSTM and Random Forest. This plot highlights that LSTM is commonly more accurate than Random Forest on the entire testing period. This is plausibly due to the ability of LSTM of learning long-term dependencies. This conclusion is further supported by the analysis of the divergent trend of the residuals on the long term. In fact, residuals of Random Forest are significantly higher than residuals of LSTM on predictions made after August 2019.

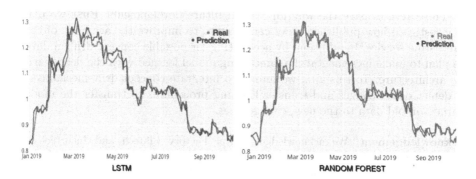

Fig. 3. Real opening prices and forecast opening prices of Exprivia S.p.a. in the considered testing period between October 12, 2018 and October 11, 2019: LSTM (left side) and Random Forest (right side).

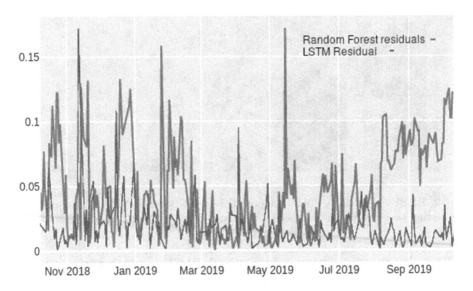

Fig. 4. Squared residuals of LSTM and Random Forest for the closing prices of Exprivia S.p.A. between October 12, 2018 and October 11, 2019.

4   Conclusion

In this paper we have illustrated the results of a preliminary investigation applying machine learning techniques to predict the closing prices of Exprivia S.p.A. in the Stock Market. This study represents the first step towards the design of a machine learning enhanced financial service for supporting the portafolio optimization. The achieved preliminary results contribute to assess the superiority of the deep learning architecture on the traditional methods in the considered task. In addition, they identify a list of ancillary ariates that aid in gaining accuracy when the problem of closing prices forecasting is addressed as a multi-variate task.

These results pave the way for several future developments. First we plan to investigate how published news can be used to improve the accuracy of this forecasting service. In addition, by accounting for possible concept drift in data, we plan to make incremental the forecasting model learned with the deep learning architecture. To this aim, we intend to integrate concept drift mechanisms to detect data changes and transfer learning procedures to transfer the model learned on old data to the new, change ones.

Acknowledgement. We acknowledge Digital Factory Fintech and Insurtech of Exprivia S.p.A for funding the work of Michele Spagnoletta. The research of Matteo Greco has been performed during his stage in Exprivia S.p.A. The research of Annalisa Appice and Donato Malerba is in partial fulfilment of theresearch objectives of the research project "Modelli e tecniche di data science per la analisi di dati strutturati" (Models and techniques of data science for the analysis of structured data) funded by University of Bari Aldo Moro.

References

1. Abu-Mostafa, Y.S., Atiya, A.F.: Introduction to financial forecasting. Appl. Intell. **6**, 205–213 (1996)
2. Asteriou, D., Hall, S.: ARIMA models and the Box-Jenkins methodology. In: Applied Econometrics, pp. 265–286. Palgrave MacMillan, London (Second ed.) (2011)
3. Chatzis, S.P., Siakoulis, V., Petropoulos, A., Stavroulakis, E., Vlachogiannakis, N.: Forecasting stock market crisis events using deep and statistical machine learning techniques. Expert Syst. Appl. **112**, 353–371 (2018). https://doi.org/10.1016/j.eswa.2018.06.032
4. Chen, T., Guestrin, C.: XGBoost: a scalable tree boosting system. In: Proceedings of the 22nd ACM SIGKDD International Conference on Knowledge Discovery and Data Mining, pp. 785–794. KDD 2016. ACM (2016). https://doi.org/10.1145/2939672.2939785
5. Ghazanfar, M.A., Alahmari, S., Aldhafiri, A., Mustaqeem, A., Maqsood, M., Azam, M.A.: Using machine learning classifiers to predict stock exchange index. Int. J. Mach. Learn. Comput. **7**, 24–29 (2017). https://doi.org/10.18178/ijmlc.2017.7.2.614
6. Granger, C.W.J.: Investigating causal relations by econometric models and cross-spectral methods. Econometrica **37**(3), 424–438 (1969). https://ideas.repec.org/a/ecm/emetrp/v37y1969i3p424-38.html
7. Hegazy, O., Soliman, O.S., Salam, M.A.: A machine learning model for stock market prediction. Int. J. Comput. Sci. Telecommun. **4**, 17–23 (2013)
8. Hochreiter, S., Schmidhuber, J.: Long short-term memory. Neural Comput. **9**(8), 1735–1780 (1997)
9. Hyndman, R., Khandakar, Y.: Automatic time series forecasting: The forecast package for R. J. Stat. Softw. **26**(3), 22 (2008). https://www.jstatsoft.org/article/view/v027i03/0
10. Louppe, G.: Understanding random forests (2015)
11. Nguyen, T.T., Yoon, S.: A novel approach to short-term stock price movement prediction using transfer learning. Appl. Sci.**9**(22), 4745 (2019). https://doi.org/10.3390/app9224745
12. Tsay, R.S.: Multivariate Time Series Analysis: With R and Financial Applications. Wiley, Hoboken (2014)

Financial Fraud Detection with Improved Neural Arithmetic Logic Units

Daniel Schlör[1]([✉]), Markus Ring[2], Anna Krause[1], and Andreas Hotho[1]

[1] University of Wuerzburg, Würzburg, Germany
{schloer,anna.krause,hotho}@informatik.uni-wuerzburg.de
[2] University of Applied Sciences and Art Coburg, Coburg, Germany
markus.ring@hs-coburg.de

Abstract. Domain specific neural network architectures have shown to improve the performance of various machine learning tasks by large margin. Financial fraud detection is such an application domain where mathematical relationships are inherently present in the data. However, this domain hasn't attracted much attention for deep learning and the design of specific neural network architectures yet. In this work, we propose a neural network architecture which incorporates recently proposed Improved Neural Arithmetic Logic Units. These units are capable of modelling mathematical relationships implicitly within a neural network. Further, inspired by a real-world credit payment application, we construct a synthetic benchmark dataset, which reflects the problem setting of automatically capturing such mathematical relations within the data. Our novel network architecture is evaluated on two real-world and two synthetic financial fraud datasets for different network parameters. We compare our proposed model with several well-established classification approaches. The results show that the proposed model is able to improve the performance of neural networks. Further, the proposed model performs among the best approaches for each dataset.

Keywords: Arithmetic neural networks · iNALU · Financial data · Fraud detection

1 Introduction

Traditional machine learning models often rely on feature engineering with expert knowledge. In contrast, one benefit of neural networks is seen in the ability of the network to find and combine relevant features. Certain network architectures have proven to be well suited for different tasks capturing the characteristics of the underlying data exceptionally well (e.g. Convolutional Neural Networks for images or Recurrent Neural Networks for sequences). However, neural network architectures specifically tailored to financial feature dependencies and mathematical relationships have not been well-studied yet.

Problem. The presence of mathematical relationships between features is a well-known fact in many financial settings [2,11]. For example, PaySim [11] is

© Springer Nature Switzerland AG 2021
V. Bitetta et al. (Eds.): MIDAS 2020, LNAI 12591, pp. 40–54, 2021.
https://doi.org/10.1007/978-3-030-66981-2_4

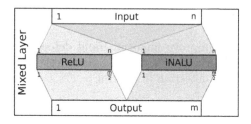

Fig. 1. Our proposed mixed layer consisting of ReLU and iNALU neurons

a mobile money simulator for fraud detection which generates bank transfers including the transmitted amount, the old account balance and new account balance as features. In this setting, these three features are highly related to each other in a mathematical sense. While neural networks are well suited for many complex data mining tasks, they often have problems with the calculation of even basic mathematical operations [18].

Objective. Although such relationships are not always directly related to the downstream-task which the machine learning model is applied to, a neural network architecture capable of capturing such relations, is able to model the data inherently better. A neural network architecture recently proposed to capture such relationships is the Improved Neural Arithmetic Unit (iNALU) [14].

In this work, we want to examine the research question if introducing iNALUs can improve the performance of neural networks on the task of financial fraud detection.

Contribution. First, we provide a short summary of existing financial fraud datasets, which are publicly available. Then, we explore the potential of enhancing neural networks using iNALUs for financial fraud detection. Since financial fraud detection is more complex than modelling mathematical relationships in the data, we propose a novel *Mixed Layer* architecture (see Fig. 1) which incorporates Rectified Linear Units (ReLUs) as general purpose neurons and iNALU neurons to capture arithmetic relationships within the data.

All approaches are evaluated on two synthetic and two real-world fraud detection datasets from the financial domain. Our findings suggest that incorporating iNALU layers significantly improves the performance on several datasets in comparison to vanilla feed forward networks with comparable network structures. Compared to several commonly used standard classifiers, our model is among the best models on each dataset and on average yields the best results.

Structure. The paper is structured as follows: The next section describes related work. Section 3 introduces the datasets used in our experiments. Our model is introduced in Sect. 4 and the experiments are described in Sect. 5. Sect. 6 discusses the results and Sect. 7 finally concludes the paper.

2   Related Work

Since this work aims to improve financial fraud detection using iNALU, this section reviews related work about modelling arithmetic relations in neural networks and financial fraud detection in general.

2.1   Modelling Arithmetic Relations

Kaiser and Sutskever [9] propose neural GPU for solving simple algorithmic tasks. Neural GPU is based on a type of Recurrent Neural Network and is able learn long binary summations and multiplications. This architecture generalizes well for long numbers, but is limited to four input symbols. Similar approaches are proposed by Kalchbrenner et al. [10] and Freivalds and Liepins [8].

Chen et al. [6] present another approach which is able to model arithmetic relations. They use reinforcement learning to solve mathematical operations such as summation, subtraction, multiplication or division. However, their approach requires the mathematical operation as an additional input to the network.

Neural Arithmetic Logic Unit (NALU) is a neural architecture designed to perform mathematical operations which is proposed by Trask et al. [18]. The NALU is not limited to certain input symbols and does not require the mathematical operation as input. The special characteristics of NALU are the restriction of the weights to the interval $[-1, 1]$ and the realization of multiplication and division in logarithmic space. The authors show in an experimental study that NALU generalizes better than traditional neurons for extrapolation tasks and achieves good results for various downstream tasks. However, NALU has some limitations as discussed in [12] and [14]. Schlör et al. [14] proposes an improved version of NALU called iNALU allowing multiplication of negative numbers and stabilization of the training.

In this work, we want to learn underlying relations within financial data implicitly. We do not want to limit ourselves to certain input symbols or explicit op-codes as input for modelling real-world datasets which generally don't fit these requirements. Therefore, we choose iNALU as basic architecture since it fits these requirements best.

2.2   Financial Fraud Detection

Financial fraud detection is very diverse and may appear in different areas such as mobile payments, credit card misuse or in ERP systems. Often, fraud represents only a very small proportion of transactions in these areas. Consequently, many financial fraud detection methods are based on anomaly detection. A comprehensive review of anomaly detection methods is given in [4]. Candola et al. [4] categorize existing approaches for anomaly detection based on their techniques and applications. Thereby, fraud detection is one of the applications being investigated. Chalapathy and Chawla [3] provide a more recent survey of deep learning for anomaly detection which among other things addresses the topic fraud detection. The authors describe existing approaches for credit card fraud detection, mobile cellular network fraud detection, insurance fraud and healthcare fraud. This survey shows that various network architectures such as Auto-Encoders, Generative Adversarial Networks, CNNs, Restricted Boltzmann Machines, or RNNs are used for fraud detection.

Baader and Krcmar [1] present an approach for fraud detection in purchase-to-pay business processes in ERP systems. The authors combine a red flag approach with process mining and try to reduce the number of false positives. Red flags are hints or indicators for fraudulent behavior and show that something irregular has happened. Process mining entails discovering, monitoring, and improving real processes through the extraction of information from event logs of IT systems. Schreyer et al. [15] use deep autoencoders to detect anomalous journal entries in ERP systems. Journal entries are standard accounting transactions which affect multiple accounts. The authors calculate an anomaly score for each journal entry which takes into account the frequency of the values and the reconstruction error of the autoencoder. Then, anomalies are detected based on a scoring systems in combination with a user-defined threshold. Schreyer et al. evaluate their work on two private data sets which encompass the entire population of journal entries of a single fiscal year.

Many works investigate the suitability of machine learning methods for credit card misuse. Wang et al. [19] evaluate Random Forest, Support Vector Machines and Capsule Networks for credit card fraud detection. The authors use a private dataset about online credit card transactions and extract several features from each transaction. The features consider customer specific historical information like the average transaction amount over the last days. Similarly, Maes et al. [13], Shen et al. [16], or Sun and Vasarhelyi [17] evaluate different neural network architectures for credit card fraud detection.

In contrast to existing approaches our work aims at evaluating the benefit of neurons which model arithmetic relations rather than achieving the best performance on the downstream tasks or specific datasets. Instead of relying on thorough feature engineering (cf. [1,19]) or unsupervised anomaly-detection methods (cf. [15]), we explicitly include neurons (iNALU) which model arithmetic relations in our neural network architectures and evaluate in a supervised classification setting. Financial data sets often cannot be shared due to privacy concerns. As a consequence, many approaches are evaluated on non-publicly available data sets and therefore lack reproducibility and the availability for other researchers to conduct follow up studies. To address this challenge, we chose four publicly available data sets in our evaluation.

3 Datasets

In this study, we use two synthetic and two real-world datasets for financial fraud detection. All datasets are highly imbalanced and contain only few fraud cases. The main dataset characteristics are summarized in Table 1.

Table 1. Main characteristics of the datasets

Dataset	Features	Samples	Benign	Fraud	Fraud-ratio	Origin
Credit	4	100 000	98 967	1 033	0.010	Synth.
PaySim	11	6 362 620	6 354 407	8 213	0.001	Synth.
CCFraud	30	284 807	284 315	492	0.002	Real
IEEE-CIS Fraud	431	590 540	569 877	20 663	0.035	Real

3.1 Credit Payment

Peer-to-Peer credits become more and more popular and open new opportunities for financial fraud. One possibility for fraud is incorrect interest-calculation. We created the synthetic credit payment dataset referred to as *Credit*, which reflects such kind of fraud.

Each data point contains the following attributes: credit sum in month x (CS_x), interest rate (IR), payment rate (PR), credit sum in month $x+1$ (CS_{x+1}), label. The mathematical relationship between the attributes is defined as follows:

$$CS_{x+1} = CS_x + \frac{CS_x \cdot IR}{12} - \lambda \cdot PR \qquad (1)$$

With a probability of 99% the credit sum is calculated correctly ($\lambda = 1$) according to Eq. 1, but with a probability of 1% we simulate a fraudulent calculation of the remaining credit by only reducing the new credit sum by 95% of the payed rate ($\lambda = 0.95$). Overall, each instance contains the columns CS_{x+1}, CS_x, IR, PR and the label *is Fraud*. For each instance the features are drawn randomly from a uniform probability distribution ($CS_x \sim \mathcal{U}(0, 10\,000)$, $IR_x \sim \mathcal{U}(0, 0.5)$, $PR_x \sim \mathcal{U}(0, 5\,000)$) with the constraint, that the credit is not overpaid i.e. $CS_{x+1} \geq 0$. In contrast to PaySim, fraudulent and benign transactions are modeled after the same probability distributions (or user-profiles) which means, the machine-learning model has to capture the mathematical relationship to predict correctly if a transaction is fraudulent. The dataset consists of 100 000 instances having 1033 fraudulent and 98 967 benign transactions. We make the dataset and the code publicly available[1].

3.2 PaySim

Paysim [11] is a multi-agent based simulator which can model financial mobile money transactions with fraudulent behavior. The simulator is based on simple mathematical statistics and is able to generate five types of transactions (*cash-in, cash-out, debit, payment* and *transfer*). From unpublished real transaction data, Lopez-Rojas et al. extracted several characteristics of the data and modelled them as user profiles. These characteristics include for example aggregated

---

[1] https://github.com/daschloer/inalu-finfraud.

transaction counts, amounts or initial account balances. Based on these user profiles, new synthetic data was generated.

Each transaction instance is described by the attributes old and new balance for both source and destination account, type of transaction, and amount of transferred money. Each row also includes a rule-based red-flag indicator for fraud, which is omitted for our experiments. We omit the account names, since they are picked randomly during the simulation without any underlying user-network structure or memory of fraudulent and non fraudulent users.

A dataset created with PaySim has been made publicly available on Kaggle[2]. The PaySim dataset contains 6 362 620 transactions of which 6 354 407 (99.871%) are benign and 8 213 (0.129%) are fraud.

3.3   Credit Card Fraud Detection

The Credit Card Fraud Detection (CCFraud) [7] dataset is a real data set of credit card transactions of European cardholders in the year 2013. The dataset contains normal and fraudulent credit card transactions. For anonymization, the authors applied principal component analysis (PCA) to all features except *time* and *amount* resulting in 28 continuous PCA transformed attributes. The feature *time* contains the seconds elapsed since the first transaction in the dataset and the feature *amount* the amount of money transferred. It contains 284 807 samples from which 492 are fraudulent.

3.4   IEEE-CIS Fraud Detection

The IEEE-CIS Credit Card Fraud Detection data set was released as part of a Kaggle[3] data science competition in 2019. This dataset is provided by Vesta Corporation and contains feature-rich representations of normal and fraudulent credit card transactions. It contains identity features such as network connection information and digital signature information about the device as well as trans-action features, e.g., product information, card information, masked undisclosed counting and match features, address and distance features as well as the amount and time-delta from a reference date. Many features are not described in detail and remain deliberately unclear for privacy and contract reasons. The dataset contains 590 540 transactions, of which 20 663 are labeled as fraud (3.5%) and 569 877 as benign (96.5%).

4   Our Model

This section first describes the iNALU architecture shortly and then proposes our novel neuronal network architecture for financial fraud detection tasks.

---

[2] https://www.kaggle.com/ntnu-testimon/paysim1.
[3] https://www.kaggle.com/c/ieee-fraud-detection/data.

4.1   INALU

Improved Neural Arithmetic Unit (iNALU) [14] is a neuron specially designed
for mathematical operations and it's architecture is shown in Fig. 2.

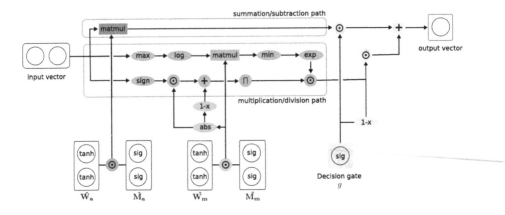

Fig. 2. iNALU architecture [14]

iNALU has one path for summation/subtraction and one path for multipli-
cation/division. A decision gate regulates the influence of both paths. Each path
has two weights matrices $\hat{W}_a$ and $\hat{M}_a$ (respectively $\hat{W}_m$ and $\hat{M}_m$). In a first
step, tanh respectively sigmoid activation functions are applied to the weight
matrices and then weight matrices are multiplied with each other element by
element. Therefore the resulting weights are scaled to the interval $[-1, 1]$ with
three plateaus at -1, 0 and $+1$. If the resulting weight is -1, iNALU performs subtractions respectively divisions. If the resulting weight is 0, iNALU
ignores the input signals. If the resulting weight is $+1$, iNALU performs summations respectively multiplications. Multiplications and divisions are calculated by
additions and subtractions of input signals in the logarithmic space. In addition,
iNALU uses max and min functions to prevent values which are too large and
calculates the resulting sign of the multiplicative path separately to allow the
multiplication and division of negative numbers. We refer to Schlör et al. [14] for
a more detailed description of iNALU and a discussion of stability and precision
in several experimental scenarios.

4.2   Improved Neural Network Architecture

Our model relies on iNALU which by design is able to model simple arithmetic
relationships. More complex relationships can be learned stacking multiple layers
of iNALUs to a deeper network. However, even in the financial domain, real-world
datasets generally contain more than mathematical relationships. Therefore, we
propose a model with each layer containing 50% general purpose non-linear
hidden units (to be precise ReLUs) and 50% iNALUs. In particular, the iNALU

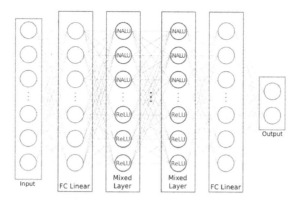

Fig. 3. Mixed Layer network model with fully connected linear input and output layers and 1 to k Mixed Layers as used in our experiments

part and the ReLU part of each layer has the full input dimension n as input, and contributes with an output dimension of $\frac{m}{2}$ concatenated to an output dimension of m for the complete layer (see Fig. 1). In combination with a linear layer as input and output layer, the network can thereby "route" and combine any input dimension to every part of each network layer by learning the weights accordingly. Therefore, the model is able to represent arithmetic as well as non-arithmetic feature relationships of varying complexity. We refer to the resulting model as shown in Fig. 3 as neuronal network with Mixed Layers (*Mixed Layers model* for short).

5 Experiments

5.1 Experimental Setup

Architecture. All experiments involve supervised training of a multi-layer-perceptron (MLP) as basic neural network architecture with a linear input, non-linear dense layers with ReLUs and a linear output layer. The number of neurons in the hidden layers varies over the experiments. To investigate the research question if introducing iNALUs can improve the performance of neural networks on the task of financial fraud detection, we use the same architecture and replace the non-linear dense layers with our Mixed Layers.

Train-Test Split. For all experiments, we use the same strategy to generate the train-test split: For training we randomly choose only few instances of the fraud class in order to keep the majority of fraudulent instances for evaluation. This approach reflects the class imbalance of the available data and emphasizes the requirement for a model to generalize from very few fraudulent samples to find new fraudulent cases when applied in a real-world scenario. In a preliminary study, we found that under-sampling the majority class to some extend didn't affect the performance negatively but reduced training-time by large margin.

Therefore only a random subset of the instances is used for training: For *Credit*, *PaySim* and *CCFraud* we use 2000 instances with a fraud-proportion of 1%, for *IEEE-CIS* we use 5000 instances with a fraud proportion of 4%. We then use the synthetic minority over-sampling technique (SMOTE) [5] to synthesize a balanced training dataset.

For the test dataset, we create a balanced split of 50% fraud and 50% benign samples, exclusively containing instances which haven't been used for training. This is motivated by the objective to study the ability of capturing mathematical relations rather than investigating effects of predicting strongly unbalanced data. To represent the variety of benign instances and avoid skewed results due to random fluctuations, we repeat this process 5 times with different random seeds and report the F1 score for the fraud class and for experiment 2 the Area Under the Receiver Operating Characteristic Curve (ROC-AUC) additionally.

Preprocessing. Each dataset is preprocessed by one-hot encoding categorical values. To assure comparability between all datasets, we follow the preprocessing strategy of the CCFraud dataset and apply PCA to all other datasets as well as min-max scaling to all datasets ensuring a valid train-test split by only fitting on training data. In a preliminary study, we verified that applying PCA to the datasets doesn't negatively impact the performance of neural networks with and without Mixed Layers. Applying PCA can also mitigate privacy issues which could possibly prevent from making a real dataset publicly available.

Training Procedure. For neural network training, we use ADAM as optimizer with a learning rate of 0.001, a weight decay of 0.0001 and Cross Entropy loss. The batch-size is set to 200 and all models are trained for 200 epochs, which have been validated as suitable training parameters in preliminary experiments.

5.2   Experiment 1

In the first experiment, we explore the influence of Mixed Layers in neural network architectures. Therefore, we construct neural networks containing Mixed Layer as well as neural networks of identical architecture exclusively with dense layers and ReLU activations. For each dataset all possible combinations between the number of input neurons (chosen from 10, 20, 30, 50 and 100) and the number of layers (chosen from 1 to 3) are evaluated to assess the performance and stability for different neural network capacities.

Results. The results are depicted in Fig. 4 and Fig. 5. The boxplots show the F1 scores on our test datasets for both neural architectures for different number of layers and varying number of hidden neurons in each layer. Each box summarizes the results of all runs for a certain parameter configuration with different random seeds. Note that both architectures share the same train- and test-splits for each run and parameter configuration. This ensures the comparability of the underlying performances and boxes for each parameter. For all dataset except IEEE-CIS, our proposed model yields very good results of F1 scores around 0.9. The performance of IEEE-CIS dataset is notably worse for both architectures

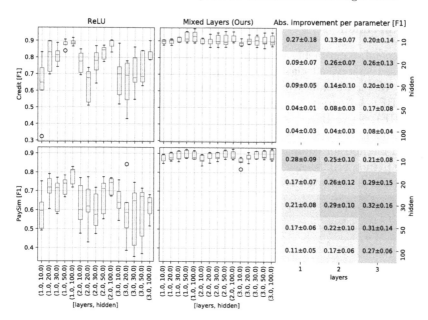

Fig. 4. F1 scores of experiment 1 on the synthetic Credit Payment and PaySim datasets. Mixed Layers describes the neural network structure as described in Sect. 4 and ReLU shows the results for the same model architecture having the Mixed Layers replaced by layers with ReLU activations. The heatmaps show the absolute improvement of Mixed Layers in comparison to the respective ReLU architecture for each parameter configuration averaged over all random seeds and their standard deviations.

and the results vary highly for different random seeds. As this dataset is much more complex regarding the number and kind of features, the task might require a different training strategy to achieve better results.

For all other datasets the performance of our model is very stable over all parameter configurations, which means that even a very small neural network consisting of one Mixed Layer and 10 hidden neurons (along with a linear input and output layer) solves the task sufficiently well. In comparison, the same architecture with one dense layer and ReLU activiations performs notably worse. To assess the actual performance improvement for each possible parameter configuration, we pairwise aligned parameters and seeds of both architectures and report the average absolute improvement per parameter configuration in Fig. 4 and Fig. 5. A positive improvement value describes a performance gain of Mixed Layers over the respective ReLU model, whereas a negative improvement means that the ReLU model performs better. The Mixed Layer model outperforms the ReLU model for all datasets except IEEE-CIS. For Credit and PaySim this observation holds for all network configurations, whereas for the CCFraud dataset large models (50 and 100 neurons) with one or two layers perform equally well.

5.3   Experiment 2

In the second experiment, we want to compare our model with several commonly used supervised classification algorithms. Precisely, we evaluate linear Support Vector Machines (SVM), Support Vector Machines with RBF kernel (SVM-RBF), k-Nearest Neighbor (kNN, $k = 3$), Decision Tree (DT), Random Forest (RF), Naïve Bayes (NB), Logistic Regression (LR) and eXtreme Gradient Boosting (XG-Boost). With Isolation Forest (IF) we also include an anomaly detection method for comparison. For all classifiers, we use the implementation from the python scikit-learn[4] library with default hyper-parameters except XG-Boost, which is evaluated using the xgboost[5] python library.

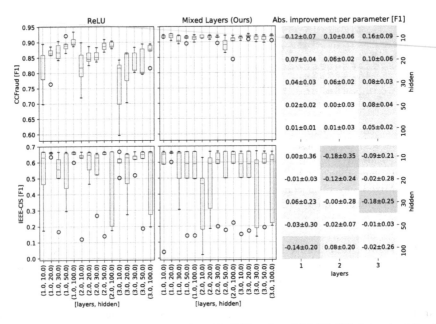

Fig. 5. F1 scores of experiment 1 on the real CCFraud and IEEE-CIS datasets. Mixed Layers describes the neural network structure as described in Sect. 4 and ReLU shows the results for the same model architecture having the Mixed Layers replaced by dense layers with ReLU activations. The heatmaps show the absolute improvement of our architecture in comparison to the respective ReLU architecture for each parameter configuration averaged over all random seeds and their standard deviations.

For the ReLU model, we use the most promising parameter configuration from experiment 1, which is one layer with 100 neurons. For the Mixed Layers model we used one layer with 20 neurons. We want to emphasize that due to the good stability of Mixed Layers over different parameter configurations, the

---

[4] https://scikit-learn.org v. 0.22.2.
[5] https://github.com/dmlc/xgboost v. 1.0.2.

Table 2. Average and standard deviation F1 score and ROC-AUC for several supervised classifiers in comparison to our model aggregated over different random seeds. For ReLU and our model, we conducted this experiment using the most promising parameter configuration from experiment 1, one layer with 100 neurons for ReLU and one layer with 20 neurons for our Mixed Layer model. The last column shows the average for each classifier over all datasets. The best results per dataset are printed in bold.

		Credit	PaySim	CCFraud	IEEE-CIS	Avg.
F1	SVM	0.88 ± 0.01	**0.89** ± 0.02	0.90 ± 0.01	0.41 ± 0.27	0.77
	SVM-RBF	**0.92** ± 0.03	0.85 ± 0.02	0.85 ± 0.07	0.43 ± 0.23	0.76
	kNN	0.89 ± 0.03	0.81 ± 0.01	0.91 ± 0.01	0.65 ± 0.03	0.81
	DT	0.87 ± 0.02	0.82 ± 0.03	0.80 ± 0.08	0.49 ± 0.04	0.74
	RF	0.89 ± 0.03	0.83 ± 0.03	0.88 ± 0.02	0.57 ± 0.05	0.80
	NB	0.85 ± 0.03	0.76 ± 0.05	0.91 ± 0.01	0.65 ± 0.02	0.80
	LR	0.88 ± 0.01	0.87 ± 0.02	**0.92** ± 0.01	0.55 ± 0.15	0.81
	XG-Boost	0.91 ± 0.02	0.84 ± 0.03	0.89 ± 0.01	0.62 ± 0.01	0.82
	IF	0.82 ± 0.01	0.81 ± 0.01	0.88 ± 0.01	**0.74** ± 0.01	0.81
	ReLU	0.89 ± 0.02	0.78 ± 0.04	0.90 ± 0.02	0.65 ± 0.03	0.81
	Mixed Layers	0.90 ± 0.01	0.88 ± 0.02	**0.92** ± 0.01	0.65 ± 0.02	**0.84**
ROC-AUC	SVM	0.95 ± 0.01	0.97 ± 0.01	0.95 ± 0.01	0.49 ± 0.12	0.84
	SVM-RBF	**0.98** ± 0.01	0.96 ± 0.01	0.93 ± 0.05	0.59 ± 0.09	0.86
	kNN	0.91 ± 0.02	0.86 ± 0.02	0.92 ± 0.01	0.54 ± 0.05	0.81
	DT	0.88 ± 0.02	0.85 ± 0.02	0.83 ± 0.05	0.52 ± 0.08	0.77
	RF	**0.98** ± 0.01	0.95 ± 0.01	**0.97** ± 0.00	0.48 ± 0.04	0.85
	NB	0.93 ± 0.02	0.92 ± 0.01	0.94 ± 0.03	0.50 ± 0.01	0.82
	LR	0.96 ± 0.01	0.96 ± 0.01	0.96 ± 0.01	0.45 ± 0.09	0.83
	XG-Boost	**0.98** ± 0.02	**0.98** ± 0.01	**0.97** ± 0.00	0.54 ± 0.01	0.87
	IF	0.95 ± 0.01	0.90 ± 0.01	0.95 ± 0.00	**0.74** ± 0.01	**0.89**
	ReLU	0.96 ± 0.01	0.88 ± 0.02	0.94 ± 0.01	0.64 ± 0.02	0.85
	Mixed Layers	0.97 ± 0.00	0.96 ± 0.01	0.96 ± 0.01	0.57 ± 0.12	0.87

parameter choice for Mixed Layer in this experiments is arbitrary and other parameter configurations perform comparably.

Results. The results of the second experiment are presented in Table 2. Our model is among the best four classifiers for all datasets. All classifiers perform well on each dataset except IEEE-CIS on which the best models except IF only achieve F1 scores of 0.65. Isolation Forest performs best on this dataset with a F1 score of 0.74 which suggests, that methods specifically tailored to anomaly detection can capture the characteristics of this dataset better. Comparing the F1 score averaged over all datasets (see Table 2, column Avg.), the Mixed Layer architecture yields significantly[6] better results compared to the best ReLU architecture. The results evaluated with the ROC-AUC metric support our findings with the exception of IEEE-CIS, where our Mixed Layers performed worse than

---

[6] $p = 0.00932$, Wilcoxon Signed-Rank Test [20] over all datasets and repetitions.

the ReLU layer. An in-depth analysis showed that two of the five repetitions with different random seeds yield notably worse results (0.41 and 0.44) which leads to the performance drop for the mean and the high standard deviation.

6 Discussion

In our experiments, we show that neural networks benefit from including Mixed Layers when applied to the task of fraud detection on four different datasets. This finding suggests that Mixed Layers improve the capability of neural networks to model mathematical relationships within the data. The neural network architectures containing our Mixed Layers have a good stability over different number of hidden neurons and layers. Even small networks yield competitive results with several well established supervised classification algorithms over different synthetic and real-world datasets. Overall the ReLU architecture seems much less stable regarding the different random seeds and parameter configurations.

Although our model was among the best classifiers for the IEEE-CIS dataset in experiment 2, all methods performed notably worse compared to the other datasets. This shows that the dataset is hard to predict in our evaluation setting and suggests that the unstable performance observed in experiment 1 is presumably not related to our model but rather to the dataset itself and might be explained by different aspects: On the one hand the dataset includes many features which might require intensive prepossessing and feature engineering. On the other hand the dataset is larger than all other datasets with respect to the number of features and fraud cases. This might require a different training procedure than the other datasets for example regarding the train-test split, the network architecture or the number of epochs for training. The observation, that Isolation Forest yields notably better results on IEEE-CIS also suggest that for more complex data sets methods adapted for anomaly detection should be used instead of standard classifiers. Since our study is not primarily conducted to achieve best performances on each dataset but rather to examine the research question, if neural network architectures benefit from iNALU neurons applied to the financial domain, our experiments focused on a fair comparison instead of thorough hyper-parameter tuning on individual datasets. However, tuning our proposed model to certain datasets in comparison to hyperparameter-tuned state-of-the-art classifiers may be interesting to investigate as future work.

On our synthetic Credit Payment dataset, some supervised classifiers performed surprisingly well, which by design of our dataset we didn't expect. Since solving the task correctly is fully dependent on capturing the correct mathematical relationship, we expected for example kNN to perform worse. We suspect the good performance is a result of applying PCA as preprocessing step, which might contribute to modeling the correct relation within the data. Examining the influence of different preprocessing steps e.g. training embedding layers end to end instead of PCA might be an interesting task for future work.

Both experiments show, that our model outperforms the respective ReLU baseline models. However, Mixed Layers with iNALUs contain more trainable parameters than linear neurons with ReLU activations. One Mixed Layer in our experiments contains 50% ReLUs and 50% iNALUs and a iNALU has 4 times more trainable parameters. For a comparison of two architectures with an equal number of trainable parameters an architecture with Mixed Layers with a hidden dimension of 20 can be compared with an architecture with ReLU Layers with a hidden dimension of 50. In this comparison (see Fig. 4 and Fig. 5) the Mixed Layers model still outperforms the ReLU based model with an F1-score of 0.837 over 0.759 averaged over all datasets.

7   Conclusion

This work examined the question if a neural network architecture which includes hidden units specifically tailored to capturing mathematical relations is beneficial for supervised classification tasks on datasets in the financial fraud domain.

We designed a new Mixed Layer for neuronal networks which incorporates iNALUs and ReLUs. Further, we constructed a synthetic benchmark dataset specifically with the difficultly of modeling a mathematical relationship, which is inspired by a real-world credit payment application. We evaluated Mixed Layer based neural networks on two real-world and two synthetic datasets and compared it to a neuronal network with ReLU activations. The experiments show that Mixed Layers are able to improve the performance of neuronal networks on financial fraud data sets. We compare our proposed model with several well-established classification approaches in a supervised evaluation setting and perform among the best approaches for each dataset.

As we designed our baseline architecture as well as our proposed model to solve a supervised task, we constructed our experiments accordingly and evaluate on a balanced dataset. However, in a real-world setting fraud and benign transaction will not occur equally frequent and the actual financial prejudice of having more false positives or more false negatives will be task depending and require a more detailed evaluation including metrics to reflect these circumstances. Moreover many approaches applied to recognize financial fraud rely on anomaly-detection or novelty-detection techniques which often use one-class or even unsupervised approaches. As we generally showed the benefit of our proposed model in the supervised setting, for future work we plan to introduce it in other evaluation settings and compare it to unsupervised approaches, as well as approaches specifically designed for anomaly or novelty-detection. Other ideas for future work include optimizing and evaluating different network architectures and to vary the proportions of iNALU and ReLU neurons in the mixed layers as well as inspecting and refining preprocessing steps.

Acknowledgements. This work was partly funded by the Federal Ministry of Education and Research of Germany as part of the DeepScan project (01IS18045A). M. R. was supported by the BayWISS Consortium Digitization.

References

1. Baader, G., Krcmar, H.: Reducing false positives in fraud detection: combining the red flag approach with process mining. Int. J. Account. Inf. Syst. **31**, 1–16 (2018)
2. Bolton, R.J., Hand, D.J.: Statistical fraud detection: a review. Stat. Sci. **17**, 235–249 (2002)
3. Chalapathy, R., Chawla, S.: Deep learning for anomaly detection: a survey. arXiv preprint arXiv:1901.03407 (2019)
4. Chandola, V., Banerjee, A., Kumar, V.: Anomaly detection: a survey. ACM Comput. Surv. (CSUR) **41**(3), 1–58 (2009)
5. Chawla, N.V., Bowyer, K.W., Hall, L.O., Kegelmeyer, W.P.: SMOTE: synthetic minority over-sampling technique. J. Artif. Intell. Res. **16**, 321–357 (2002)
6. Chen, K., Dong, Y., Qiu, X., Chen, Z.: Neural arithmetic expression calculator. arXiv preprint arXiv:1809.08590 (2018)
7. Dal Pozzolo, A., Caelen, O., Johnson, R.A., Bontempi, G.: Calibrating probability with undersampling for unbalanced classification. In: IEEE Symposium Series on Computational Intelligence, pp. 159–166. IEEE (2015)
8. Freivalds, K., Liepins, R.: Improving the neural GPU architecture for algorithm learning. In: Workshop on Neural Abstract Machines & Program Induction (NAMPI), International Conference on Machine Learning (ICML) (2018)
9. Kaiser, L., Sutskever, I.: Neural GPUs learn algorithms. In: International Conference on Learning Representations (2016)
10. Kalchbrenner, N., Danihelka, I., Graves, A.: Grid long short-term memory. arXiv preprint arXiv:1507.01526 (2015)
11. Lopez-Rojas, E., Elmir, A., Axelsson, S.: PaySim: a financial mobile money simulator for fraud detection. In: European Modeling and Simulation Symposium (EMSS), pp. 249–255. Dime University of Genoa (2016)
12. Madsen, A., Rosenberg Johansen, A.: Measuring Arithmetic Extrapolation Performance (2019)
13. Maes, S., Tuyls, K., Vanschoenwinkel, B., Manderick, B.: Credit card fraud detection using Bayesian and neural networks. In: International Naiso Congress on Neuro Fuzzy Technologies, pp. 261–270 (2002)
14. Schlör, D., Ring, M., Hotho, A.: iNALU: improved neural arithmetic logic unit. arXiv e-prints arXiv:2003.07629, March 2020
15. Schreyer, M., Sattarov, T., Borth, D., Dengel, A., Reimer, B.: Detection of anomalies in large scale accounting data using deep autoencoder neural networks. In: GPU Technology Conference - Silicon Valley (2018)
16. Shen, A., Tong, R., Deng, Y.: Application of classification models on credit card fraud detection. In: International Conference on Service Systems and Service Management, pp. 1–4. IEEE (2007)
17. Sun, T., Vasarhelyi, M.A.: Predicting credit card delinquencies: an application of deep neural networks. Intell. Syst. Account. Finance Manag. **25**(4), 174–189 (2018)
18. Trask, A., Hill, F., Reed, S.E., Rae, J., Dyer, C., Blunsom, P.: Neural arithmetic logic units. In: Advances in Neural Information Processing Systems, pp. 8035–8044 (2018)
19. Wang, S., Liu, G., Li, Z., Xuan, S., Yan, C., Jiang, C.: Credit card fraud detection using capsule network. In: IEEE International Conference on Systems, Man, and Cybernetics (SMC), pp. 3679–3684 (2018)
20. Wilcoxon, F.: Individual comparisons by ranking methods. Biometrics Bull. **1**(6), 80–83 (1945)

Information Extraction From the GDELT Database to Analyse EU Sovereign Bond Markets

Sergio Consoli[1(✉)], Luca Tiozzo Pezzoli[1], and Elisa Tosetti[1,2]

[1] Joint Research Centre, Directorate A-Strategy, Work Programme and Resources,
Scientific Development Unit, European Commission,
Via E. Fermi 2749, 21027 Ispra, VA, Italy
{sergio.consoli,luca.tiozzo-pezzoli}@ec.europa.eu
[2] Department of Management, Universitá Ca' Foscari Venezia,
Cannaregio 873, 30121 Fondamenta San Giobbe, Venice, Italy
elisa.tosetti@unive.it

Abstract. In this contribution we provide an overview of a currently on-going project related to the development of a methodology for building economic and financial indicators capturing investor's emotions and topics popularity which are useful to analyse the sovereign bond markets of countries in the EU.These alternative indicators are obtained from the Global Data on Events, Location, and Tone (GDELT) database, which is a real-time, open-source, large-scale repository of global human society for open research which monitors worlds broadcast, print, and web news, creating a free open platform for computing on the entire world's media. After providing an overview of the method under development, some preliminary findings related to the use case of Italy are also given. The use case reveals initial good performance of our methodology for the forecasting of the Italian sovereign bond market using the information extracted from GDELT and a deep Long Short-Term Memory Network opportunely trained and validated with a rolling window approach to best accounting for non-linearities in the data.

Keywords: Big data · Government yield spread · GDELT · Machine learning · Features engineering

1 Introduction and Preliminaries

Economic and fiscal policies conceived by international organizations, governments, and central banks heavily depend on economic forecasts, in particular during times of economic turmoil like the one we have recently experienced with the COVID-19 virus spreading world-wide [30]. The accuracy of economic forecasting and nowcasting models is however still problematic since modern economies are subject to numerous shocks that make the forecasting and nowcasting tasks extremely hard, both in the short and in the medium-long run.

V. Bitetta et al. (Eds.): MIDAS 2020, LNAI 12591, pp. 55–67, 2021.
https://doi.org/10.1007/978-3-030-66981-2_5

In this context, the use of recent *Big Data* technologies for improving forecasting and nowcasting for several types of economic and financial applications has high potentials. In a currently on-going project we are designing a methodology to extract alternative economic and financial indicators capturing investor's emotions, topics popularity, and economic and political events, from the *Global Database of Events, Language and Tone (GDELT)*[1] [17], a novel big database of news information. GDELT is a real-time, open-source, large-scale repository of global human society for open research which monitors worlds broadcast, print, and web news. The news-based economic and financial indicators extracted from GDELT can be used as alternative features to enrich forecasting and nowcasting models for the analysis of the sovereign bond markets of countries in the EU.

The very large dimensions of GDELT make unfeasible the use of any relational database and require ad-hoc big data management solutions to perform any kind of analysis in reasonable time. In our case, after GDELT data are crawled from the Web by means of custom REST APIs[2], we use *Elasticsearch* [13,24] to host and interact with the data. Elasticsearch is a popular and efficient NO-SQL big data management system whose search engine relies on the Lucene library[3] to efficiently transform, store, and query the data.

After GDELT data are stored into our Elasticsearch infrastructure, a feature selection procedure selects the variables having higher forecasting potentials to analyse the sovereign bond market of the EU country under study. The selected variables capture, among others, investor's emotions, economic and political events, and popularity of news thematics for that country. These additional variables are included into economic forecasting and nowcasting models with the goal of improving their performance. In current research we are experimenting different models, ranging from traditional economic models to novel machine learning approaches, like Gradient Boosting Machines and Recurrent Neural Networks (RNNs), which have been shown to be successful in various forecasting problems in Economics and Finance (see e.g. [4,6–8,16,18,29] among others).

2   Related Work

The recent surge in the government yield spreads in countries within the Euro area has originated an intense debate about the determinants and sources of risk of sovereign spreads. Traditionally, factors such as the creditworthiness, the sovereign bond liquidity risk, and global risk aversion have been identified as the main factors having an impact on government yield spreads [3,22]. However, a recent literature has pointed at the important role of financial investor's sentiment in anticipating interest rates dynamics [19,26]. An early paper that has used a sentiment variable calculated on news articles from the Wall Street Journal is [26]. In this work it is showed that high levels of pessimism are a relevant predictor of convergence of the stock prices towards their fundamental

---

[1] GDELT website: https://blog.gdeltproject.org/.
[2] See https://blog.gdeltproject.org/gdelt-2-0-our-global-world-in-realtime/.
[3] https://lucene.apache.org/.

values. Other recent works in finance exist on the use of emotions extracted from social media, financial microblogs, and news to improve predictions of the stock market (e.g. [1,9]). In the macroeconomics literature, [14] has looked at the informational content of the Federal Reserve statements and the guidance that these statements provide about the future evolution of monetary policy. Other papers ([27,28] and [25] among others) have used Latent Dirichlet allocation (LDA) to classify articles in topics and to extract a signal with predictive power for measures of economic activity, such as GDP, unemployment and inflation [12]. These results, among others, have shown the high potentials of the information extracted from news variables on monitoring and improving the forecasts of the business cycle [9].

Machine learning approaches in the existing literature for controlling financial indexes measuring credit risk, liquidity risk and risk aversion include the works in [3,5,10,11,20], among others. Several efforts to make machine learning models accepted within the economic modeling space have increased exponentially in recent years (see e.g.. [4,6–8,16,18,29] among others).

3   GDELT Data

GDELT analyses over 88 million articles a year and more than 150,000 news outlets. Its dimension is around 8 TB, growing 2TB each year [17]. For our study we rely on the "Global Knowledge Graph (GKG)" repository of GDELT, which captures people, organizations, quotes, locations, themes, and emotions associated with events happening in print and web news across the world in more than 65 languages and translated in English. Themes are mapped into commonly used practitioners' topical taxonomies, such as the "World Bank (WB) Topical Ontology"[4]. GDELT also measures thousands of emotional dimensions expressed by means of, e.g., the "Harvard IV-4 Psychosocial Dictionary"[5], the "WordNet-Affect dictionary"[6], and the "Loughran and McDonald Sentiment Word Lists dictionary"[7], among others. For our application we use the GDELT GKG fields from the World Bank Topical Ontology (i.e. WB themes), all emotional dimensions (GCAM), and the name of the journal outlets.

The huge number of unstructured documents coming from GDELT are re-engineered and stored on an ad-hoc Elasticsearch infrastructure [13,24]. Elasticsearch is a popular and efficient document-store built on the Apache Lucene search library[8] and providing real-time search and analytics for different types of complex data structures, like text, numerical data, or geospatial data, that have been serialized as JSON documents. Elasticsearch can efficiently store and

---

[4] https://vocabulary.worldbank.org/taxonomy.html.

[5] Harvard IV-4 Psychosocial Dictionary: http://www.wjh.harvard.edu/~inquirer/homecat.htm.

[6] WordNet-Affect dictionary: http://wndomains.fbk.eu/wnaffect.html.

[7] Loughran and McDonald Sentiment Word Lists: https://sraf.nd.edu/textual-analysis/resources/.

[8] https://lucene.apache.org/.

index data in a way that supports fast searches, allowing data retrieval and aggregate information functionalities via simple REST APIs to discover trends and patterns in the stored data.

4   Feature Selection

We use the available World Bank Topical Ontology to understand the primary focus (theme) of each article and select the relevant news whose main themes are related to events concerning bond market investors. Hence, we select only articles such that the topics extracted by GDELT fall into one of the following WB themes of interest: *Macroeconomic Vulnerability and Debt*, and *Macroeconomic and Structural Policies*. To make sure that the main focus of the article is one of the selected WB topics, we retain only news that contain in their text at least three keywords belonging to these themes. The aim is to select news that focus on topics relevant to the bond market, while excluding news that only briefly report macroeconomic, debt and structural policies issues. We consider only articles that are at least 100 words long. From the large amount of information selected, we construct features counting the total number of words belonging to all WB themes and GCAMs detected each day. We also create the variables "Number of mentions" denoting the word count of each location mentioned in the selected news. We further filter the data by using domain knowledge to retain a subset of GCAM dictionaries that qualitatively may have potentials to our analysis. Then we retain only the variables having a standard deviation calculated over the full sample greater than 5 words and allowing a 10% of missing values on the total number of days. Finally we perform a correlation analysis across the selected variables, normalized by number of daily articles. If the correlation between any two features is above 80% we give preference to the variable with less missing values, while if the number of missing values is identical and the two variables belong to the same category (i.e. both are themes or GCAMs), we randomly pick one of them. Finally, if the number of missing values is identical but the two variables belong to the same category, we consider the following order of priority: GCAM, WB themes, GDELT themes, locations.

5   Preliminary Results

Here we show some preliminary results on the application of the described methodology for the use case of Italy. The main objective of this empirical exercise is to assess the predictive power of GDELT selected features for the forecasting of the Italian sovereign bond market.

We have extracted data from Bloomberg on the term-structure of government bond yields for Italy over the period 2 March 2015 to 31 August 2019. We have calculated the sovereign spread for Italy against Germany as the difference between the Italian 10 year maturity bond yield minus the German counterpart. We have also extracted the standard level, slope and curvature factors of the term-structure using the Nelson and Siegel [23] procedure and included these classical factors into the model. Being the government bond yields a highly persistent and non-stationary process, we have considered its log-differences and obtained a stationary series of daily changes representing our prediction target, illustrated in Fig. 1. This kind of forecasting exercise is an extremely challenging task, as the target series behaves similarly to a random walk process. Missing data, related to weekends and holidays, have been dropped from the target time series, giving a final number of 468 data points.

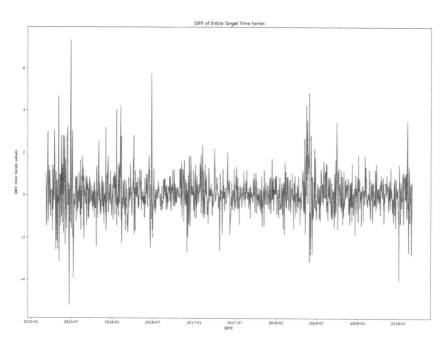

Fig. 1. Log-differences of the sovereign spread for Italy against Germany as the difference between the Italian 10 year maturity bond yield minus the German counterpart.

For our Italian case study, we have also extracted the news information from GKG in GDELT from a set of around 20 newspapers for Italy, published over the considered period of the analysis. After this selection procedure we obtained a total of 18,986 articles, with a total of 2,978 GCAM, 1,996 Themes and 155 locations. Applying the feature selection procedure described above, we have extracted 31 dimensions of the General Inquirer Harvard IV psychosocial Dictionary, 61 dimensions of Roget's Thesaurus, 7 dimensions of the Martindale Regressive Imagery and 3 dimensions of the Affective Norms for English Words (ANEW) dictionary. After the features engineering procedure, we have been left with a total of 45 variables, of which 9 are themes, 34 are GCAM, 2 locations. The selected topics contained WB themes such as Inflation, Government, Central Banks, Taxation and Policy, which are indeed important thematics discussed in the news when considering interest rates issues. Moreover, selected GCAM features included optimism, pessimism or arousal, which explore the emotional state of the market. Figure 2 shows the top correlated covariates with respect to the target.

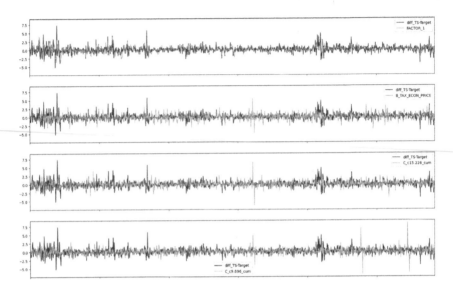

Fig. 2. Log-differences of the sovereign spread for Italy against Germany as the difference between the Italian 10 year maturity bond yield minus the German counterpart.

Several studies in the literature have shown that during stressed periods, complex non-linear relationships among explanatory variables affect the behaviour of the output target which simple linear models are not able to capture. For this reason, in this empirical exercise we have used a deep Long Short-Term Memory Network (LSTM) [15] to best accounting for non-linearities and assessing the predictive power of the selected GDELT variables. The LSTM was implemented relying on the DeepAR model available in Gluon Time Series (GluonTS) [2][9], an open-source library for probabilistic time series modelling that focuses on deep learning-based approaches and interfacing Apache MXNet[10]. DeepAR is an LSTM model working into a probabilistic setting, that is, predictions are not restricted to point forecasts only, but probabilistic forecastings are produced according to a user-defined predictive distribution (in our case a student t-distribution was experimentally selected). For our experiment we have set experimentally to use 2 RNN layers, each having 40 LSTM cells, and used a learning rate equal to 0.001. The number of training epochs was set to 500, with training loss being the negative log-likelihood function.

---

[9] Available at: https://gluon-ts.mxnet.io/#gluonts-probabilistic-time-series-modeling.
[10] Available at: https://mxnet.apache.org/.

We have used a robust scaling for the training variables by adopting statistics robust to the presence of outliers. That is, we have removed the median to each time series, and the data were scaled according to the interquartile range. Furthermore we have adopted a rolling window estimation technique where the first estimation sample started at the beginning of March and ended in May 2017. For each window, one step-ahead forecasts have been calculated. The whole experiment required to run few hours in parallel on 40 cores at 2.10 GHz each into an Intel(R) Xeon(R) E7 64-bit server having overall 1 TB of shared RAM.

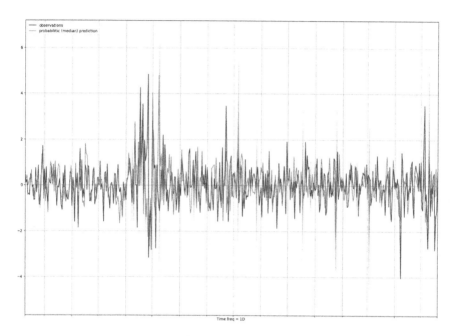

Fig. 3. Median forecasts (green) and observations for the target series (blue) for the entire forecasting period. (Color figure online)

Figure 3 shows the observations for the target time series (blue line) together with the median forecast (dark green line) and the confidence interval in lighter green. To better visualize the differences between observed and predicted time series, we have reported the same plot on a smaller time range (50 days) in Figure 4. A qualitative analysis of the figure suggests that the forecasting model does a reasonable job at capturing the variability and volatility of the time series.

We have also computed a number of commonly used evaluation metrics [21], such as the mean absolute scaled error (MASE), the symmetric mean absolute percentage error (sMAPE), the root mean square error (RMSE), and the (weighted) quantile losses (wQuantileLoss), that is the quantile negative log-likelihood loss weighted with the density. The obtained in-sample and out-of-sample results are shown in Table 1. As expected the results worsen passing

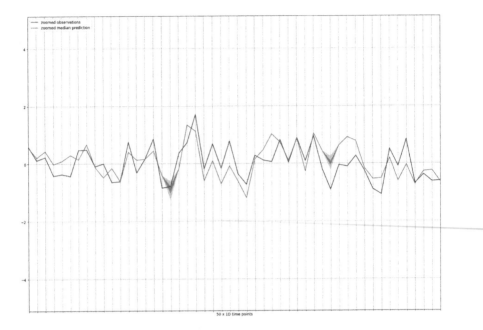

Fig. 4. Probabilistic forecasts (green) and observations for the target series (blue) for the first 50 days in the testing period. The green continuous line shows the median of the probabilistic predictions, while the lighter green areas represents an higher confidence interval. (Color figure online)

Table 1. Forecasting results of the LSTM model in terms of MASE, sMAPE, RMSE, and wQuantileLoss error metrics.

Metrics	LSTM results	
	In-sample	Out-of-sample
MASE	0.112	0.682
sMAPE	0.130	1.148
RMSE	0.493	0.885
wQuantileLoss[0.1]	0.050	0.869
wQuantileLoss[0.3]	0.115	0.899
wQuantileLoss[0.5]	0.151	0.914
wQuantileLoss[0.7]	0.121	0.923
wQuantileLoss[0.9]	0.047	0.907

from the in-sample to the out-of-sample setting, but the gap is absolutely acceptable, confirming a good generalization capability of the trained LSTM model. The model showed higher performance at high (0.9) and low (0.1) quantiles with lower weighted quantile losses. Figure 5 illustrates the median absolute forecast error (MAFE, in orange) against the real time series observations (in blue).

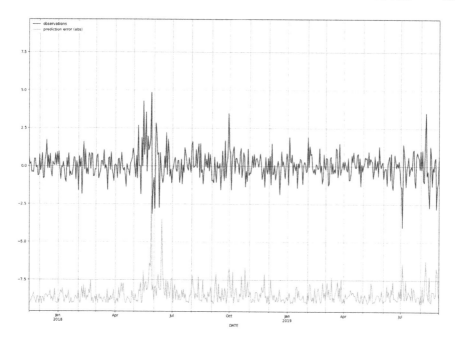

Fig. 5. Mean absolute forecast error (MAFE) (orange) against real observations (blue). (Color figure online)

The performance of the model slightly worsen from the end of May to July 2018, corresponding to a period of political turmoil in Italy. Indeed, on the 29th of May, the Italian spread sharply rose reaching 250 basis point. Investors where particularly worried about the possibility of anti-euro government and not confident on the formation of a stable government. From June until November 2018, a series of discussions about deficit spending engagements and possible conflicts with European fiscal rules continued to worry the markets. The spread strongly increased in October and November with values around 300 basis point. We can see this also from the performance of our model which worsen a bit in this stressed period, which however the model looks to handle quite well anyway. Since 2019, the Italian political situation started to improve and the spread smoothly declined, especially after the agreement with Brussels on budget deficit in December 2018. However, some events hit the Italian economy afterwards, such as the EU negative outlook and the European parliament elections which contributed to a temporary increase on interest rates. Our model performs quite well in this period in terms of absolute error ratios showing a good robustness.

Figure 6 shows a scatter plot amongst the median out-of-sample forecasted points and the real observations. To some degree the points in the scatter plot roughly follow the diagonal, showing a fine correlation among the forecasted points and the real observations, and suggesting good quality of the forecasting results. This is also confirmed by the acceptable value of 0.23 computed for the

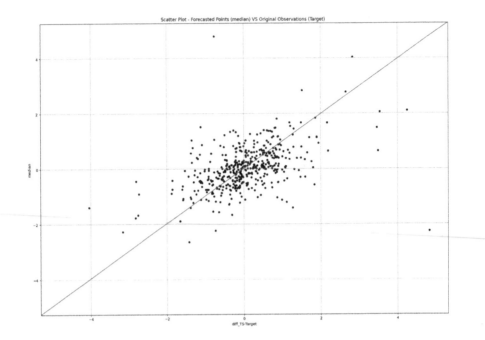

Fig. 6. Scatter plot amongst the median out-of-sample forecasted points and the real observations.

R-squared metrics on the out-of-sample median forecasts for such a challenging prediction exercise. This value of the R-squared measure indicates that the LSTM model explains a quite ample variability of the response data around its median, suggesting a certain degree of closeness among the forecasted data and the real observations.

6   Conclusion and Overlook

In this contribution we have presented our work-in-progress related to the development of a methodology for building alternative economic and financial indicators capturing investor's emotions and topics popularity from GDELT, the Global Data on Events, Location, and Tone database, a free open platform containing real-time worlds broadcast, print, and web news. The currently on-going project in which this work is developed is aimed at producing improved forecasting methods to analyse the sovereign bond markets of countries in the EU. We have reported some preliminary results on the application of this methodology for predicting the Italian sovereign bond market. This use case reveals initial good performance of the methodology, suggesting the validity of the approach. Using the information extracted from the Italian news media contained in GDELT combined with a deep Long Short-Term Memory Network opportunely trained and validated with a rolling window approach, we have been able to obtain quite good forecasting results.

This work represents one of the first to study the behaviour of government yield spreads and financial portfolio decisions in the presence of classical yield curve factors and information extracted from news. We believe that these new measures are able to capture and predict changes in interest rates dynamics especially in period of turmoil. Overall, the paper shows how to use a large scale database as GDELT to derive financial indicators in order to capture future intentions of agents in sovereign bond markets.

Certainly more research is still needed to be exploited in the directions of the presented work. First we will try to improve the performance of the implemented DeepAR model by tweaking architecture and optimizing the hyperparameters of the LSTM model. Furthermore, in current research we are experimenting other different prediction models, ranging from traditional economic methods to other novel machine learning approaches, including Gradient Boosting Machines and neural forecasting methods. In a future extended version of the paper we will compare and thoroughly analyze the performance of these methods to better exploit the non-linear effects of the dependent variables. Interpretability of the implemented machine learning models by using, e.g., computed Shapley values, will be an important object of future investigation in order to finely assess the contributions of the different covariates in the models predictions.

Acknowledgments. The authors would like to thank the colleagues of the Centre for Advanced Studies at the Joint Research Centre of the European Commission for helpful guidance and support during the development of this research work.

References

1. Agrawal, S., Azar, P., Lo, A.W., Singh, T.: Momentum, mean-reversion and social media: evidence from StockTwits and Twitter. J. Portfolio Manag. **44**, 85–95 (2018)
2. Alexandrov, A., et al.: GluonTS: probabilistic time series models in Python. CoRR, abs/1906.05264 (2019). http://arxiv.org/abs/1906.05264
3. Beber, A., Brandt, M.W., Kavajecz, K.A.: Flight-to-quality or flight-to-liquidity? Evidence from the Euro-area bond market. Rev. Financ. Stud. **22**(3), 925–957 (2009)
4. Benidis, K., et al.: Neural forecasting: introduction and literature overview. CoRR, abs/2004.10240 (2020). https://arxiv.org/abs/2004.10240
5. Bernal, O., Gnabo, J.-Y., Guilmin, G.: Economic policy uncertainty and risk spillover in the Eurozone. J. Int. Money Finance **65**(C), 24–45 (2016)
6. Borovykh, A., Bohte, S., Oosterlee, C.W.: Conditional time series forecasting with convolutional neural networks. Lecture Notes in Computer Science (including subseries Lecture Notes in Artificial Intelligence and Lecture Notes in Bioinformatics), vol. 10614, pp. 729–730 (2017)
7. Chang, Y.-C., Chang, K.-H., Wu, G.-J.: Application of eXtreme gradient boosting trees in the construction of credit risk assessment models for financial institutions. Appl. Soft Comput. J. **73**, 914–920 (2018)
8. Deng, S., Wang, C., Wang, M., Sun, Z.: A gradient boosting decision tree approach for insider trading identification: an empirical model evaluation of china stock market. Appl. Soft Comput. J. **83** (2019)

9. Dridi, A., Atzeni, M., Reforgiato Recupero, D.: FineNews: fine-grained semantic sentiment analysis on financial microblogs and news. Int. J. Mach. Learn. Cybern., 1–9 (2018)

10. Favero, C., Pagano, M., von Thadden, E.-L.: How does liquidity affect government bond yields? J. Financ. Quant. Anal. **45**(1), 107–134 (2010)

11. Garcia, A.J., Gimeno, R.: Flight-to-liquidity flows in the Euro area sovereign debt crisis. Technical report, Banco de Espana Working Papers (2014)

12. Gentzkow, M., Kelly, B., Taddy, M.: Text as data. J. Econ. Lit. (2019, to appear)

13. Gormley, C., Tong, Z.: Elasticsearch: The Definitive Guide. O' Reilly Media, Sebastopol (2015)

14. Hansen, S., McMahon, M.: Shocking language: understanding the macroeconomic effects of central bank communication. J. Int. Econ. **99**, S114–S133 (2016)

15. Hochreiter, S., Schmidhuber, J.: Long short-term memory. Neural Comput. **9**, 1735–1780 (1997)

16. Koenecke, A., Gajewar, A.: Curriculum learning in deep neural networks for financial forecasting. In: Bitetta, V., Bordino, I., Ferretti, A., Gullo, F., Pascolutti, S., Ponti, G. (eds.) MIDAS 2019. LNCS (LNAI), vol. 11985, pp. 16–31. Springer, Cham (2020). https://doi.org/10.1007/978-3-030-37720-5_2

17. Leetaru, K., Schrodt, P.A.: GDELT: global data on events, location and tone, 1979–2012. Technical report, KOF Working Papers (2013)

18. Liu, J., Wu, C., Li, Y.: Improving financial distress prediction using financial network-based information and GA-based gradient boosting method. Comput. Econ. **53**(2), 851–872 (2019). https://doi.org/10.1007/s10614-017-9768-3

19. Loughran, T., McDonald, B.: When is a liability not a liability? Textual analysis, dictionaries and 10-ks. J. Finance **66**(1), 35–65 (2011)

20. Manganelli, S., Wolswijk, G.: What drives spreads in the Euro area government bond markets? Econ. Policy **24**(58), 191–240 (2009)

21. Mehdiyev, N., Enke, D., Fettke, P., Loos, P.: Evaluating forecasting methods by considering different accuracy measures. Procedia Comput. Sci. **95**, 264–271 (2016)

22. Monfort, A., Renne, J.-P.: Decomposing Euro-area sovereign spreads: credit and liquidity risks. Rev. Finance **18**(6), 2103–2151 (2013)

23. Nelson, C., Siegel, A.F.: Parsimonious modeling of yield curves. J. Bus. **60**(4), 473–489 (1987)

24. Shah, N., Willick, D., Mago, V.: A framework for social media data analytics using Elasticsearch and Kibana. Wireless Networks (2018, in press)

25. Shapiro, A.H., Sudhof, M., Wilson, D.: Measuring news sentiment. Federal Reserve Bank of San Francisco Working Paper (2018)

26. Tetlock, P.C.: Giving content to investor sentiment: the role of media in the stock market. J. Finance **62**(3), 1139–1168 (2007)

27. Thorsrud, L.A.: Nowcasting using news topics. big data versus big bank. Norges Bank Working Paper (2016)

28. Thorsrud, L.A.: Words are the new numbers: a newsy coincident index of the business cycle. J. Bus. Econ. Stat., 1–17 (2018)

29. Yang, X., He, J., Lin, H., Zhang, Y.: Boosting exponential gradient strategy for online portfolio selection: an aggregating experts' advice method. Comput. Econ. **55**(1), 231–251 (2020). https://doi.org/10.1007/s10614-019-09890-2

30. Zhang, D., Hu, M., Ji, Q.: Financial markets under the global pandemic of COVID-19. Finance Res. Lett., 101528 (2020)

Multi-objective Particle Swarm Optimization for Feature Selection in Credit Scoring

Nikita Kozodoi[1,2(✉)] and Stefan Lessmann[1]

[1] Humboldt University of Berlin, Berlin, Germany
nikita.kozodoi@hu-berlin.de
[2] Monedo, Hamburg, Germany

Abstract. Credit scoring refers to the use of statistical models to support loan approval decisions. An ever-increasing availability of data on potential borrowers emphasizes the importance of feature selection for scoring models. Traditionally, feature selection has been viewed as a single-objective task. Recent research demonstrates the effectiveness of multi-objective approaches. We propose a novel multi-objective feature selection framework for credit scoring that extends previous work by taking into account data acquisition costs and employing a state-of-the-art particle swarm optimization algorithm. Our framework optimizes three fitness functions: the number of features, data acquisition costs and the AUC. Experiments on nine credit scoring data sets demonstrate a highly competitive performance of the proposed framework.

Keywords: Credit scoring · Feature selection · Multi-objective optimization

1 Introduction

Financial institutions use credit scoring models to support loan approval decisions [10]. Due to the unprecedented availability of data on potential credit applicants and growing access of financial institutions to new data sources, the data used to train scoring models tend to be high-dimensional [6].

Feature selection aims at removing irrelevant features to improve the model performance, which is traditionally considered as a single-objective task [7]. In credit scoring, feature selection can be treated as a multi-objective problem with multiple goals. In addition to optimizing model performance, companies strive to reduce the number of features as public discourse and regulatory requirements are calling for comprehensible credit scoring models [9]. Furthermore, financial institutions often purchase data from external providers such as credit bureaus and banks in groups of features. This creates a need for a separate account for the data acquisition costs [13]. The conflicting nature of these objectives motivates us to consider feature selection as a multi-objective optimization problem.

© Springer Nature Switzerland AG 2021
V. Bitetta et al. (Eds.): MIDAS 2020, LNAI 12591, pp. 68–76, 2021.
https://doi.org/10.1007/978-3-030-66981-2_6

Recent research has demonstrated the effectiveness of multi-objective feature selection in credit scoring using genetic algorithms (GA) [9]. However, previous studies have not considered a simultaneous optimization of the number and the cost of features. Moreover, techniques based on particle swarm optimization (PSO) have been recently shown to outperform GAs in other domains [15].

This paper proposes a multi-objective feature selection framework for credit scoring and makes two contributions. First, our framework addresses three distinct objectives: (i) the number of features, (ii) the cost of features and (iii) the model performance. Optimizing both the number and the cost of features is crucial as purchasing data from multiple sources introduces grouping that decorrelates the two objectives. Increasing the number of data providers incurs higher costs, whereas adding individual features does not affect costs if the new features are purchased in a group with the already included features. Second, we adjust a state-of-the-art PSO algorithm denoted as AgMOPSO to the feature selection context to improve the performance of the search algorithm.

2   Related Work

Standard techniques consider feature selection as a single-objective task [7]. Recent studies have shown the importance of accounting for multiple objectives such as model performance and cardinality of the feature set [3,9]. A simple approach to account for multiple objectives is to aggregate them into one objective or introduce optimization constraints to a single-objective task. This requires additional information such as weights of objectives or budget conditions. In contrast, a truly multi-objective approach produces a set of non-dominated solutions – a Pareto frontier – where improving one objective is impossible without worsening at least one of the others. Provided with such a frontier, a decision-maker can examine the trade-off between the objectives depending on the context.

Previous research on multi-objective feature selection has employed evolutionary algorithms such as GA and PSO. Most studies use a non-dominated sorting genetic algorithm NSGA-II, which is a well-known algorithm for multi-objective optimization [8]. NSGA-III was proposed as a successor of NSGA-II to handle challenges of many-objective optimization [3]. The usage of GAs has also been recently challenged by the proposal of PSO techniques that demonstrate a superior performance [15,16]. Research outside of the feature selection domain has also suggested other optimization techniques such as Gaussian processes [4].

Credit scoring is characterized by multiple conflicting objectives, such as model performance and comprehensibility [9,13]. Nevertheless, research on multi-objective feature selection in credit scoring has been scarce. Maldonado et al. use support vector machines (SVM) to optimize performance while minimizing feature costs as a regularization penalty [12,13]. Both methods are embedded in SVMs and output a single solution instead of a Pareto frontier. Kozodoi et al. argue for a model-agnostic multi-objective approach [9]. Their framework uses NSGA-II and is limited to two objectives, assuming the number of features to be indicative of both model comprehensibility and data acquisition costs.

We extend the previous work on feature selection in credit scoring by adapting a state-of-the-art PSO algorithm to perform the feature search and considering feature costs as a distinct objective. A common practice of purchasing data in groups of features reduces the correlation between the objectives and provides an opportunity for multi-criteria optimization. The number of features serves as a proxy for model comprehensibility and interpretability, whereas feature costs indicate the data acquisition costs faced by a financial institution.

3   Proposed Framework

We propose a multi-objective feature selection framework based on the external archive-guided MOPSO algorithm (AgMOPSO), which demonstrates superior performance compared to GAs in various optimization tasks [17]. PSO is a meta-heuristic method that solves an optimization problem by evolving a population of candidate solutions (i.e., particles) that iteratively navigate through the search space. AgMOPSO encompasses two stages: initialization and feature search.

The algorithm initializes with a random swarm of n particles. Each particle is represented by a real-valued vector of length k, where k is the number of features. The particle values are restricted to $[0, 1]$ and indicate the probability of a feature being selected. The initialization is followed by an iterative process of guiding the swarm towards new solutions using the immune-based and PSO-based search and evolving an archive that stores the non-dominated solutions.

The immune-based search produces new particles by applying genetic operators such as cloning, crossover and mutation to the existing particles [17]. The PSO-based search creates new particles by updating the values of each particle using a decomposition-based approach. The particle position in the k-dimensional feature space is adjusted by moving it towards the swarm leaders, i.e., particles that perform best in each of the three objectives [1].

After each search round, we evaluate solutions. We train a model that includes features corresponding to the rounded particle values and evaluate three fitness functions: (i) the number of features, (ii) feature acquisition costs and (iii) the model performance in terms of the area under the ROC curve (AUC). Based on the evaluated fitness, we store non-dominated solutions representing different feature subsets in the archive. Until the maximum size of the archive is reached, all non-dominated solutions are added to the archive. Once the archive is full, we calculate the crowding degree of new particles that indicates the density of surrounding solutions. If the new solution displays a better crowding degree than at least one archive solution, it replaces the most crowded solution in the archive.

4   Experimental Setup

4.1   Data

Table 1 displays the data sets used in the experiments. All data sets come from a retail credit scoring context. Data sets *australian* and *german* are part of the UCI

Table 1. Credit scoring data sets

Data set	Sample size	No. features	Default rate
australian	690	42	.44
german	1,000	61	.30
Thomas	1,125	28	.26
hmeq	5,960	20	.20
cashbus	15,000	1,308	.10
lendingclub	43,344	206	.07
pakdd2010	50,000	373	.26
paipaidai	60,000	1,934	.07
gmsc	150,000	68	.07

Repository. Data sets *thomas* and *hmeq* are provided by [14] and [2]; *paipaidai* is collected from [11]. Data sets *pakdd*, *lendingclub* and *gmsc* are provided for data mining competitions on PAKDD and Kaggle platforms.

Each data set contains a binary target variable indicating whether a customer has repaid a loan and a set of features describing characteristics of the applicant, the loan and, in some cases, the applicant's previous loans. As illustrated in Table 1, the sample size and the number of features vary across the data sets, which allows us to test our feature selection framework in different conditions.

4.2 Experimental Setup

We consider a multi-criteria feature selection problem with three objectives: (i) the number of selected features, (ii) feature acquisition costs, (iii) the AUC. Each of the nine data sets is randomly partitioned into training (70%) and holdout sets (30%). We perform feature selection with AgMOPSO within four-fold cross-validation on the training set. The performance of the selected feature subsets is evaluated on the holdout set. To ensure robustness, the performance is aggregated over 20 modeling trials with different random data partitioning.

Since data on feature acquisition costs are not available in all data sets, we simulate costs similar to [15]. The cost of each feature is drawn from a Uniform distribution in the interval [0, 1]. To simulate feature groups, we introduce a cost-based grouping for categorical features. Each categorical feature is transformed with dummy encoding. Next, we assign acquisition costs to dummy features: if one dummy stemming from a specific categorical feature is selected, other dummies from the same feature can be included at no additional cost.

NSGA-II and NSGA-III with the same three objectives as AgMOPSO serve as benchmarks. The meta-parameters of the algorithms are tuned using grid search on a subset of training data. To ensure a fair comparison, the population size and the number of generations for NSGA-II and NSGA-III are set to the same values as for the AgMOPSO. We also use a full model with all features as a benchmark. L2-regularized logistic regression serves as a base classifier.

We use five evaluation metrics common in multi-objective optimization to reflect different characteristics of the evolved solution frontiers: (i) hypervolume (HV) is an overall performance metric that indicates the objective space covered by the solutions; (ii) overall non-dominated vector generation (ONVG) is the number of distinct non-dominated solutions; (iii) two-set coverage (TSC) reflects the convergence of the frontier; (iv) spacing (SPC) considers how evenly the solutions are distributed; (v) maximum spread (SPR) accounts for the solution spread. The calculation of the SPC and the SPR requires knowledge of the true Pareto frontier, which is difficult to estimate due to the high dimensionality of the feature space. We combine the non-dominated solutions across the three algorithms and 20 trials to form an adequate approximation of the true frontier.

We also compare the full model with the single best-performing solutions of the evolutionary algorithms that achieve the highest AUC compared to the other solutions on the evolved Pareto frontiers. The single solutions are compared in terms of the three considered objectives.

5   Results

Table 2 provides the experimental results. For each data set, we rank algorithms in the five multi-objective optimization metrics and report the mean ranks across the 20 trials. We also report the mean AUC, data acquisition costs and the number of features of the single solutions with the highest AUC.

Overall, AgMOPSO outperforms the GA-based benchmarks in three performance metrics, achieving the lowest average rank in the ONVG, the TSC and the HV. According to the Nemenyi test [5], differences in algorithm ranks are significant at a 5% level. The superior performance of AgMOPSO is mainly attributed to a higher cardinality and a better convergence of the evolved frontier compared to NSGA-II and NSGA-III. This is indicated by the best performance of AgMOPSO in the ONVG and the TSC on seven out of nine data sets.

In terms of the diversity of the evolved frontier, AgMOPSO does not outperform the benchmarks. In the feature selection application, diversity of solutions is of secondary importance as we are mainly interested in models from a subspace that covers the best-performing models. A more relevant indicator and an apparent strength of AgMOPSO compared to the competing algorithms is its better convergence to the true frontier, which indicates that it misses a smaller number of non-dominated solutions in the relevant search space.

The mean correlation between the number of features and data acquisition costs across the non-dominated solutions is .7455. This emphasizes the importance of a separate optimization of the two feature-based objectives.

Due to a large number of noisy features, the performance of the full model is suboptimal. Comparing the full model to the solutions of multi-objective algorithms with the highest AUC, we see that all three evolutionary algorithms identify a feature subset that achieves a higher AUC, lower costs and a smaller number of features on eight data sets. On average, AgMOPSO reduces data acquisition costs and the number of selected features by 58.77% and 76.20%,

Table 2. Comparing performance of feature selection methods

Data set	Algorithm	ONVG	TSC	SPC	SPR	HV	AUC	FC	NF
australian	AgMOPSO	**1.70**	**1.80**	2.25	**1.45**	**1.80**	.9184	1.96	9.35
	NSGA-II	2.05	1.93	2.00	2.10	2.15	.9170	1.95	8.15
	NSGA-III	2.10	2.05	**1.75**	2.15	2.05	.9162	**1.49**	**6.90**
	Full model					–	.6751	6.81	42
german	AgMOPSO	**1.20**	1.87	2.55	**1.35**	1.95	.7823	3.78	15.70
	NSGA-II	1.75	**1.82**	2.15	1.50	**1.40**	**.7824**	3.24	13.65
	NSGA-III	2.90	2.07	**1.30**	2.70	2.65	.7723	**1.95**	**9.05**
	Full model					–	.5806	9.97	61
thomas	AgMOPSO	1.65	**1.75**	2.05	**1.75**	1.95	.6368	1.47	4.50
	NSGA-II	1.70	**1.75**	2.05	1.80	**1.80**	.6363	1.36	3.70
	NSGA-III	**1.55**	**1.75**	**1.50**	1.95	1.85	.6364	**1.00**	**3.30**
	Full model					–	**.6375**	6.81	28
hmeq	AgMOPSO	**1.20**	1.77	2.45	**1.45**	**1.45**	**.7660**	3.23	9.45
	NSGA-II	1.95	1.88	2.15	1.75	2.25	.7651	3.14	9.10
	NSGA-III	2.60	2.02	**1.40**	2.60	2.30	.7604	**2.32**	**5.85**
	Full model					–	.5730	5.76	20
cashbus	AgMOPSO	**1.85**	**1.72**	1.70	2.25	2.30	.6450	**3.52**	**10.55**
	NSGA-II	1.95	2.00	2.90	**1.05**	2.30	.6426	14.64	44.95
	NSGA-III	2.15	2.08	**1.40**	2.70	**1.40**	**.6606**	8.22	27.50
	Full model					–	.5233	321.10	1308
lendingclub	AgMOPSO	**1.60**	1.95	1.90	1.70	2.05	.6169	1.96	20.25
	NSGA-II	1.80	2.12	2.30	**1.50**	2.60	.6149	2.08	18.60
	NSGA-III	2.40	**1.75**	**1.80**	2.25	**1.35**	**.6176**	**1.81**	**16.75**
	Full model					–	.5725	6.24	206
pakdd2010	AgMOPSO	**1.45**	**1.75**	2.15	2.05	**1.00**	**.6254**	8.10	115.55
	NSGA-II	1.95	2.28	2.75	**1.00**	2.00	.6252	9.47	252.05
	NSGA-III	2.55	1.95	**1.10**	2.95	3.00	.6220	**7.65**	**69.65**
	Full model					–	.5748	14.58	373
paipaidai	AgMOPSO	2.85	**1.80**	1.70	2.50	1.85	.6639	5.84	**36.15**
	NSGA-II	1.60	2.03	2.85	**1.00**	**1.40**	.6727	8.88	245.40
	NSGA-III	**1.50**	1.92	**1.45**	2.40	2.75	**.6802**	**5.82**	76.35
	Full model					–	.4956	57.89	1934
gmsc	AgMOPSO	**1.00**	**1.72**	2.45	**1.50**	**1.35**	**.8603**	4.72	25.50
	NSGA-II	1.95	1.90	2.20	1.70	1.90	.8602	4.71	23.15
	NSGA-III	3.00	2.30	**1.35**	2.65	2.75	.8556	**3.52**	**15.75**
	Full model					–	.6437	4.85	68

Note: ONVG = overall non-dominated vector generation, TSC = two-set coverage, SPC = spacing, SPR = maximum spread, HV = hypervolume, AUC = area under the ROC curve, FC = feature acquisition costs, NF = number of selected features.

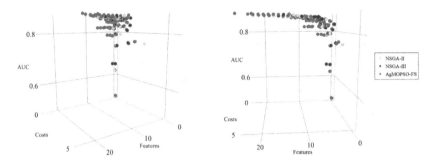

Fig. 1. Pareto frontiers for *gmsc* depicted from two angles.

respectively. AgMOPSO also outperforms GA-based benchmarks in terms of the AUC on five data sets. At the same time, solutions of NSGA-III entail lower costs and a smaller number of features compared to AgMOPSO, which comes at the cost of a lower average AUC. These results indicate that AgMOPSO more effectively explores regions of the search space associated with a higher AUC.

Figure 1 illustrates the Pareto frontiers outputted by the three feature selection algorithms on one of the trials on *gmsc*. The figure demonstrates the ability of AgMOPSO to identify a larger number of non-dominated solutions in the search subspace with a high AUC compared to the GA-based benchmarks.

6   Discussion and Future Work

This paper proposes a multi-objective framework for feature selection in credit scoring using the AgMOPSO algorithm. We perform feature selection using three fitness functions reflecting relevant credit scoring objectives: the number of features, data acquisition costs, and model performance. The performance of our framework is assessed on nine real-world credit scoring data sets.

The results suggest that AgMOPSO is a highly competitive multi-objective feature selection framework, as indicated by standard quality criteria for multi-objective optimization. Compared to other evolutionary algorithms, AgMOPSO more effectively explores regions of the search space associated with a high model performance, while also substantially reducing the number of features and the data acquisition costs compared to a model using all features.

In future studies, we plan to conduct a more in-depth analysis of AgMO-POSO. It would be interesting to compare results with the solutions evolved by two-objective feature selection algorithms that ignore data acquisition costs. Analysis of the impact of correlation between the objectives on the algorithm performance could also shed more light on conditions in which the number and the cost of features should be considered as separate objectives. In addition, computing the running times and the number of generations before convergence would contribute a new angle to compare feature selection algorithms.

AgMOPOSO has a wide set of meta-parameters, which poses an opportunity for a systematic sensitivity analysis that could provide deeper insights into the appropriate values. For instance, the diversity of the evolved solutions could be improved by adjusting the crossover and mutation operations within the search. Using different base learners could help evaluate gains given a model with a built-in feature selection mechanism (e.g., L1-regularized logistic regression).

Our multi-objective feature selection framework could be extended to other application areas, such as fraud detection or churn prediction. For both of these applications, customer data are typically gathered from different sources and therefore provides opportunities for group-based cost optimization.

References

1. Al Moubayed, N., Petrovski, A., McCall, J.: D2MOPSO: MOPSO based on decomposition and dominance with archiving using crowding distance in objective and solution spaces. Evol. Comput. **22**(1), 47–77 (2014)
2. Baesens, B., Roesch, D., Scheule, H.: Credit Risk Analytics: Measurement Techniques, Applications, and Examples in SAS. Wiley, New York (2016)
3. Bidgoli, A.A., Ebrahimpour-Komleh, H., Rahnamayan, S.: A many-objective feature selection algorithm for multi-label classification based on computational complexity of features. In: 2019 14th International Conference on Computer Science and Education (ICCSE), pp. 85–91. IEEE (2019)
4. Bradford, E., Schweidtmann, A.M., Lapkin, A.: Efficient multiobjective optimization employing Gaussian processes, spectral sampling and a genetic algorithm. J. Global Optim. **71**(2), 407–438 (2018). https://doi.org/10.1007/s10898-018-0609-2
5. Demšar, J.: Statistical comparisons of classifiers over multiple data sets. J. Mach. Learn. Res. **7**, 1–30 (2006)
6. Gambacorta, L., Huang, Y., Qiu, H., Wang, J.: How do machine learning and non-traditional data affect credit scoring? New evidence from a Chinese fintech firm. In: Working paper, Bank for International Settlements (2019)
7. Guyon, I., Gunn, S., Nikravesh, M., Zadeh, L.A.: Feature Extraction: Foundations and Applications, vol. 207. Springer, Heidelberg (2008). https://doi.org/10.1007/978-3-540-35488-8
8. Hamdani, T.M., Won, J.-M., Alimi, A.M., Karray, F.: Multi-objective feature selection with NSGA II. In: Beliczynski, B., Dzielinski, A., Iwanowski, M., Ribeiro, B. (eds.) ICANNGA 2007, Part I. LNCS, vol. 4431, pp. 240–247. Springer, Heidelberg (2007). https://doi.org/10.1007/978-3-540-71618-1_27
9. Kozodoi, N., Lessmann, S., Papakonstantinou, K., Gatsoulis, Y., Baesens, B.: A multi-objective approach for profit-driven feature selection in credit scoring. Decis. Support Syst. **120**, 106–117 (2019)
10. Lessmann, S., Baesens, B., Seow, H.V., Thomas, L.C.: Benchmarking state-of-the-art classification algorithms for credit scoring: an update of research. Eur. J. Oper. Res. **247**(1), 124–136 (2015)
11. Li, H., Zhang, Y., Zhang, N.: Evaluating the well-qualified borrowers from PaiPaiDai. Procedia Comput. Sci. **122**, 775–779 (2017)
12. Maldonado, S., Flores, Á., Verbraken, T., Baesens, B., Weber, R.: Profit-based feature selection using support vector machines - general framework and an application for customer retention. Appl. Soft Comput. **35**, 740–748 (2015)
13. Maldonado, S., Pérez, J., Bravo, C.: Cost-based feature selection for support vector machines: an application in credit scoring. Eur. J. Oper. Res. **261**(2), 656–665 (2017)

14. Thomas, L., Crook, J., Edelman, D.: Credit scoring and its applications. In: SIAM (2017)
15. Zhang, Y., Gong, D.W., Cheng, J.: Multi-objective particle swarm optimization approach for cost-based feature selection in classification. IEEE/ACM Trans. Comput. Biol. Bioinf. **14**(1), 64–75 (2015)
16. Zhang, Y., Gong, D.W., Sun, X.Y., Guo, Y.N.: A PSO-based multi-objective multi-label feature selection method in classification. Sci. Rep. **7**(1), 1–12 (2017)
17. Zhu, Q., et al.: An external archive-guided multiobjective particle swarm optimization algorithm. IEEE Trans. Cybern. **47**(9), 2794–2808 (2017)

A Comparative Analysis of Temporal Long Text Similarity: Application to Financial Documents

Vipula Rawte[1]([×]) [ID], Aparna Gupta[2] [ID], and Mohammed J. Zaki[1] [ID]

[1] Computer Science Department, Rensselaer Polytechnic Institute, Troy, NY, USA
rawtev@rpi.edu, zaki@cs.rpi.edu
[2] Lally School of Management, Rensselaer Polytechnic Institute, Troy, NY, USA
guptaa@rpi.edu

Abstract. Temporal text documents exist in many real-world domains. These may span over long periods of time during which there tend to be many variations in the text. In particular, variations or the similarities in a pair of documents over two consecutive years could be meaningful. Most of the textual analysis work like text classification focuses on the entire text snippet as a data instance. It is therefore important to study such similarities besides the entire text document. In Natural Language Processing (NLP), the task of textual similarity is important for search and query retrieval. This task is also better known as Semantic Textual Similarity (STS) that aims to capture the semantics of two texts while comparing them. Also, state-of-the-art methods predominantly target short texts. Thus, measuring the semantic similarity between a pair of long texts is still a challenge. In this paper, we compare different text matching methods for the documents over two consecutive years. We focus on their similarities for our comparative analysis and evaluation of financial documents, namely public 10-K filings to the SEC (Securities and Exchange Commission). We further perform textual regression analysis on six quantitative bank variables including *Return on Assets (ROA)*, *Earnings per Share (EPS)*, *Tobin's Q Ratio*, *Tier 1 Capital Ratio*, *Leverage Ratio*, and *Z-score*, and show that textual features can be effective in predicting these variables.

Keywords: Temporal changes · Text similarity · Long text regression · SEC 10-K reports

1 Introduction

In general, over long periods of time, the language tends to evolve [8]. Specifically, the written language present in the form of unstructured text documents changes a lot given the amount of text data generated daily. Studying how the texts change over time is an interesting area. A lot of text mining research is being done on the entire text documents or short texts, such as tweets, messages,

V. Bitetta et al. (Eds.): MIDAS 2020, LNAI 12591, pp. 77–91, 2021.
https://doi.org/10.1007/978-3-030-66981-2_7

reviews, and so on. Some works also target temporal text mining [13] and temporal networks [35]. Temporal text documents exist in real-world domains like clinical texts, news articles, financial statements, and restaurant and product reviews. For this work, we focus on the financial disclosure statements, i.e., 10-K documents from the Securities and Exchange Commission (SEC) in the finance domain. In particular, we look at the Sects. 7 and 7A: "Management's Discussion and Analysis of financial conditions and results of operations" (MD&A). The reason to choose the finance domain and Sect. 7/7A in 10-K documents is threefold:

1. Sect. 7A contains forward-looking statements about the company, which report the company's operations and financial results. It may also talk about the potential risks.
2. The length of Sect. 7 is longer than an average text document, the approximate longest section being an average of ~40000 words as reported in Table 2.
3. The nature of the MD&A section makes it suitable to find if any relationship exists between the textual similarities and bank financial performance variables [6,19].

Measuring the similarity between texts is a challenging task in NLP because of linguistic, semantic and knowledge-based factors. There exist a number of methods ranging from traditional count based methods to neural networks and knowledge based methods for measuring textual similarity. The basic idea is to map the texts onto a vector space model (VSM) such that there are two vectors for a pair of documents. A cosine similarity metric is used to measure the similarity between these two vectors by computing the dot product between two such normalized vectors. Textual similarity has a wide range of applications in real-world domains such as legal [2,22,31,33], academic [20,21], finance [6, 30], and medicine [36,38]. It can also be useful for several NLP tasks such as document classification, clustering, and retrieval.

Contributions: In this study, we compare different document representation techniques and similarities between two temporal text documents over two consecutive years and apply them to the section 7/7A of public 10-K filings. Given its significance and applications in NLP, we choose several state-of-the-art methods in our comparative study. We further use these similarities to predict six bank variables using linear regression. We also study some interesting patterns between the textual similarities and the numeric values.

2   Related Work

Document dating [12,34], Neural Networks (NN) based diachronic framework for temporal text classification [11], and dynamic topic modeling [4,24,37] are a few of the recent works in the area of temporal text document analysis.

In [6], the authors show how the modification score varies with the length of the document over years. This score is computed by using the cosine similarity between the document vectors constructed using the term frequency-inverse document frequency (tf-idf) scores. In more recent work [5], the authors create a multi-dimensional measure of financial statement peer-to-peer similarity which is purely quantitative.

In the field of Information Retrieval (IR), document similarity is widely applicable for tasks for as document clustering, document retrieval, query search, document deduplication, question answering, and so on. In [32], a comparative study of various textual similarity methods such as tf-idf (and other related extensions), topic models (Latent Semantic Indexing (LSI)) [7], and neural models (paragraph vector, doc2vec [18]) is presented. It highlights that tf-idf still remains a good option when it comes to long text documents as compared to the other complex methods.

A lot of work on short text similarity exists with some standard evaluation benchmark datasets such as SemEval STS [1]. [15] shows how word embeddings (vector representations of words) constructed from unlabelled data can be used to compute the semantic similarity by finding the cosine similarity between two vectors. This includes the semantics instead of just lexical or syntactic information. In [29], a NN-based technique is described to measure the textual similarity. The authors used a Siamese Convolutional Neural Network (CNN) to capture the relevance of a word in a sentence in order to create a word representation. Then they used a Siamese LSTM (Long-Short Term Memory) to analyze a pair of sentences, and their similarity is computed using the Manhattan distance. [27] proposed a novel way to compute similarity based on the present term set in order to tackle the issue where several documents have an identical degree of similarity to a specific document. This measure is based on the term weights and the number of terms that exist in at least one of the two documents. External knowledge and word embeddings based method to measure semantic similarity is proposed in [26].

For longer documents, [21] proposed a novel joint word-embedding model based semantic matching of long documents by incorporating domain-specific semantics information into the basic context of the words. This model is then applied to academic documents by including semantic profiles for research purpose, methodology, and domain to create the embedding. Since the lengths of two documents can vary, representing long documents as vectors is not always helpful. Thus, it is important to address the length difference between two documents when computing their similarity. The method in [9] shows how to represent a longer document as a representation of the latent topics, and the shorter document as just an average of the word embedding it is composed of. Finally, they used cosine similarity to measure the document similarity. They also showed the ineffectiveness of doc2vec and Word Mover's Distance [17] in their document similarity task. In [25], the authors present a simple, unsupervised method for pairwise document matching. They try to improve [9] by first averaging over the word embeddings, and then compute the cosine similarity between these two averages. In [3], the authors proposed a knowledge-based semantic textual similarity technique called Context Semantic Analysis (CSA) which relies on a RDF knowledge base such as DBpedia and Wikidata to extract a Semantic Context Vector to represent a document.

3   Methodology

We try to closely follow the work in [6] where the authors define a *rawscore* to measure the changes between two documents. In this work, we instead look at the similarities between two documents. For this purpose, we examine some text-based methods to measure similarity between two financial documents. We further categorize these methods into the following three groups.

3.1   Bag-of-Words (BOW)

It represents text in a vector form as the occurrence of words in a given document.

1. **Count Vectorizer:** This is a classical method to create a document vector, which is solely based on the word counts of the words present in the corpus vocabulary. Such a document vector gives more importance to the most frequent words even if they are not relevant.
2. **tf-idf:** The above issue is overcome by weighting the word scores using a formula given in Eq. 1 to construct a document vector.

$$tf\text{-}idf_{i,j} = tf_{i,j} \times \log\left(\frac{N}{df_i}\right) \tag{1}$$

where, $tf_{i,j}$ is the number of occurrences of word i in document j, df_i is the number of documents containing word i, and N is the total number of documents.

3.2   Word/Document Embedding

The BOW methods do not capture any semantics of the text and so we study some additional methods below.

1. **tf-idf + Latent Semantic Analysis (LSA):** Unlike bag-of-words or neural network models, LSA works on the core idea of document-term matrix and Singular Value Decomposition (SVD) as a dimensionality reduction technique for the sparse document-term matrix. It decomposes the entire matrix into a separate document-topic matrix and a topic-term matrix. Thus, it is able to capture the semantics from the textual documents. It uses the basic idea of SVD on the document-term matrix by diving it into three matrices U, Σ and V so that: $X = U\Sigma V^t$ where U is the terms-topics matrix, Σ is the topics importance matrix and V^t is the topics-documents matrix.

2. **doc2vec:** This is an unsupervised document embedding technique to create document vectors [18]. It is analogous to the word2vec [23] technique to create word vectors. Similar to word2vec, doc2vec has two variants, distributed memory (DM) and distributed bag of words (DBOW). We use the *gensim* library[1] to create the doc2vec representations. For training, we set the parameters as *epochs=100 and window_size=15*. Next, we use the following three different word embedding techniques in the doc2vec model.

   word2vec: It captures the local context of a given word within a window size to create word vectors [23].

   GloVe: It captures both local and global context of a given word by constructing the co-occurrence matrix of the words in a document [28].

   fasttext: This model breaks down an out-of-vocabulary word into sub-words and has improvements over word2vec and GloVe in some NLP tasks [14]. For doc2vec, we use both the variants, DM and DBOW. These methods have shown improvements in results on tasks such semantic textual similarity, but may perform worse when there are misspellings of the same word. This issue of misspelling is handled well in string matching techniques, discussed below, where a word is reconstructed from the original word using the minimum edit distance.

3.3 String Matching

The simplest way to compare two strings is with a measurement of edit distance required to reconstruct a string from the original string.

1. **Fuzzywuzzy:** Fuzzy string matching technique compares two strings to find matches where there are misspellings or just partial words. It is called *fuzzy* because it uses an 'approximate' string matching technique based on Levenshtein Distance to calculate the edit distance, using the formula given in Eq. 2. We use a fast, optimized Python library Fuzzywuzzy[2] for the task of string matching.

$$\text{lev}_{a,b}(i,j) = \begin{cases} \max(i,j) & \text{if } \min(i,j)=0 \\ \min \begin{cases} \text{lev}_{a,b}(i-1,j)+1 \\ \text{lev}_{a,b}(i,j-1)+1 \\ \text{lev}_{a,b}(i-1,j-1)+1_{(a_i \neq b_j)} \end{cases} & \text{otherwise} \end{cases} \quad (2)$$

where $1_{(a_i \neq b_j)}$ denotes 0 when $a = b$ and 1 otherwise. Finally, the *Levenshtein similarity ratio* is computed based on the Levenshtein distance, and is calculated using the formula in Eq. 3.

---

[1] https://radimrehurek.com/gensim/models/doc2vec.html.
[2] https://github.com/seatgeek/fuzzywuzzy.

$$\frac{(|a| + |b|) - \text{lev}_{a,b}(i,j)}{|a| + |b|} \tag{3}$$

where $|a|$ and $|b|$ are the lengths of sequence a and sequence b, respectively.

3.4   Similarity Metrics

For two given documents, we compare them to measure the similarity between them. A similarity metric is thus used to quantify such similarity between two texts. Having looked at the three groups of approaches for representing documents, we next consider some common similarity metrics based on different input representations.

1. **Jaccard Similarity:** It is used to compute similarity between two strings which are represented as sets. The formula to calculate Jaccard similarity is given in Eq. 4.

$$J(A, B) = \frac{|A \cap B|}{|A \cup B|} = \frac{|A \cap B|}{|A| + |B| - |A \cap B|} \tag{4}$$

where A and B are two sets containing tokens.

2. **Cosine similarity:** It calculates similarity between two vectors by measuring the cosine of the angle between them and it is given by formula in Eq. 5. In our case, we use techniques like tf-idf, tf-idf+LSA, word2vec and doc2vec to represent each document as a vector. We then use cosine similarity to measure the similarity between these two vectors.

$$\text{cosine\_sim}(A, B) = cos(\theta) = \frac{A \cdot B}{\|A\| \times \|B\|} = \frac{\sum_{i=1}^{n} A_i \times B_i}{\sqrt{\sum_{i=1}^{n} A_i^2} \times \sqrt{\sum_{i=1}^{n} B_i^2}} \tag{5}$$

where, A_i and B_i are the ith components of the vectors A and B, respectively, and θ is the angle between A and B in the n-dimensional vector space.

3. **Pearson Correlation Coefficient:** It is a statistical measure that finds the linear correlation between two vectors A and B. It ranges between -1 and 1, where -1 is total negative linear correlation, 0 is no linear correlation, and 1 is total positive linear correlation. The formula to calculate the correlation is given in Eq. 6.

$$\text{PCC}(A, B) = \frac{\text{cov}(A, B)}{\sigma_A \sigma_B} = \frac{\sum_{i=1}^{n} (A_i - \bar{A})(B_i - \bar{B})}{\sqrt{\sum_{i=1}^{n} (A_i - \bar{A})^2 (B_i - \bar{B})^2}} \tag{6}$$

where $\text{cov}(A, B)$ is the covariance, σ_A is the standard deviation of A and σ_B is the standard deviation of B.

4 Experiments and Results

4.1 Dataset and Preprocessing

For our experiments, we use the dataset described in [30], which contains the 10-K filings for US banks for the period between 2006 and 2016. We only use Sects. 7 and 7A in our experiments. We report all the dataset statistics in Table 1. For preprocessing, we use *nltk* word tokenizer[3] to tokenize the documents. We then use *beautifulsoup4*[4] to remove all HTML tags. We also remove all the stopwords and punctuation (except., %, and $), and convert the numerals and email address into #. Finally, we use *nltk PorterStemmer* to stem the tokens. We use *scikit-learn*[5] to compute tf-idf scores. We obtain the bank variable values from the dataset used in [10]. Also, we use *sklearn MinMaxScaler* to scale all the values for our evaluation. We show the distribution of all six bank variables in Fig. 1 and their statistics in Table 2. Code and data of our approach is available at https://github.com/vr25/temp_change_text_reg.

Table 1. Dataset size

	# Documents
Total	5321
After extracting items 7, 7A	3396
Data pairs for two consecutive years	2337

Table 2. Data statistics: the min is 0 and max is 1 after scaling.

	ROA	EPS	Tobin's Q ratio	Tier 1 cap ratio	Leverage ratio	Z-score
Mean	7.84E-01	7.50E-01	4.50E-01	2.44E-01	4.40E-01	9.46E-02
Std. dev.	5.27E-02	5.92E-02	7.42E-02	5.14E-02	3.23E-02	1.07E-01

It is interesting to note that we also observe that the average length of Sect. 7, 7A increases over time as shown in Fig. 2. Finally, we study the effect of document length on textual similarity in Fig. 3. We also plot the difference in document lengths (denoted delta), i.e., $year2$ - $year1$ lengths, and thus we can get negative values when $year1$ document is longer than $year2$ document. We also find that the similarity increases with the increase in document length beyond 40000 words.

---

[3] https://www.nltk.org/.
[4] https://pypi.org/project/beautifulsoup4/.
[5] https://scikit-learn.org/.

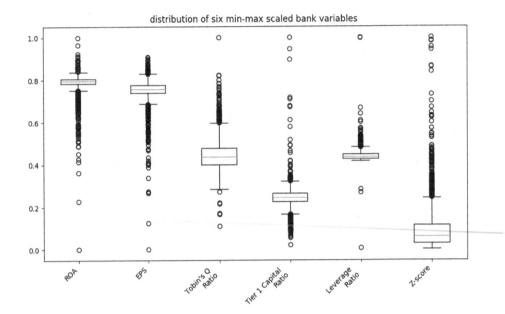

Fig. 1. This box-whisker plot shows the distribution of all six bank variables.

4.2 Evaluation Metric

In order to evaluate our experiments on text regression task, we use mean squared error (mse) (also known as standard deviation of residuals) metric given in the following Eq. 7:

$$\text{mean squared error (mse)} = \frac{1}{N} \sum_{i=1}^{N} (y_i - \hat{y}_i)^2 \tag{7}$$

4.3 Results and Discussion

All the experiments were performed on a machine with a 2.3 GHz Intel Xeon Processor E5-2670 v3 Processor, with 251G RAM and 80 CPU cores and Python 3.6 environment.

We study and compare some standard document embedding techniques and document similarity measures. We use these methods on textual data available from a dataset of Sects. 7 and 7A from 10-K filings. Our main goal here is to examine how the textual similarities from two consecutive years perform when compared to the entire document. We also compare the textual features with the numeric scores.

Regression: For evaluation, we perform linear regression and ridge regression. We use linear regression where the independent variable (numeric score) is the similarity measure and the dependent variable (numeric score) is one of the

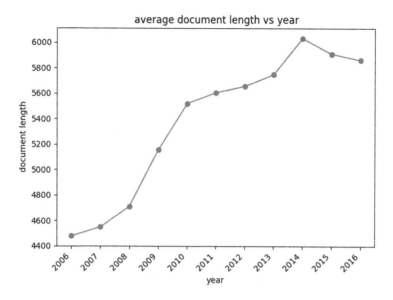

Fig. 2. Document length vs. year pairs

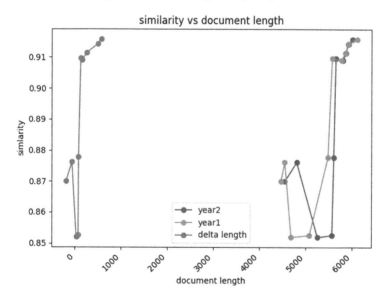

Fig. 3. Textual similarity (cosine) [doc2vec (word2vec)] vs. document length

six bank variables (done independently). Next, we also, compare these textual similarities with the entire textual content. We do this by selecting text from (1) *year1*, (2) *year2*, (3) concatenation of *year1* and *year2* document vectors (using tf-idf method), and (4) combination of *year1* and *year2* documents. In (3), we first compute the tf-idf based vector representation of the text for two

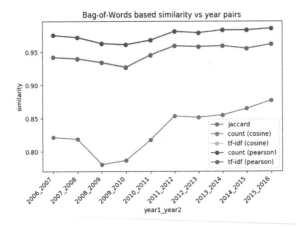

Fig. 4. This plot shows Bag-of-Words similarity across year pairs.

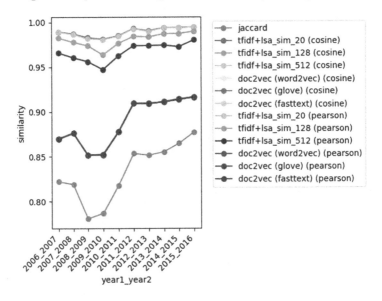

Fig. 5. This plot shows document vector based similarity across year pairs.

consecutive years *year1* and *year2* and then concatenate these vectors, whereas in (4) we first merge the textual strings from *year1* and *year2* and then create a tf-idf based vector for the combined text. We now perform ridge regression on these textual features. For both, linear and ridge regression, we then perform a 10-fold cross validation, and compute the mean squared error.

Baseline: We use *year1* scores to predict *year2* scores for all six bank variables. Thus, we are using historical values to predict the future values as a numeric baseline which is similar to the baseline proposed in [16] for textual regression to predict future stock volatility. Similarly, we also use the delta (*year2 - year1*) values of these scores to predict the *year2* values.

Comparative Analysis: We categorize the jaccard similarity metric into the Bag-of-Words type since this similarity measure uses sets of words which have no specific word order. We also use count-based and tf-idf based methods to construct document vectors. We study the relation between these methods across pairs of years between 2007 and 2016 in Fig. 4. We observe a slight drop in similarity between years 2008 and 2010, during the previous financial crisis. In Fig. 4 and Fig. 5, since we get identical results for the cosine similarity and PCC these plots overlap each other.

For embedding based methods, we choose tf-idf + LSA (with different number of topics) and doc2vec (with different word embedding techniques). For two given vectors, we compute (1) cosine similarity and (2) PCC (Pearson Correlation Coefficient), as shown in Fig. 5. Interestingly, we find that jaccard smilarity (bag-of-words) and doc2vec (document embedding) show similar trend across year pairs and this pattern is quite similar to tf-idf+lsa methods too.

Bag-of-Words: For BOW methods, we compute similarity between two tokenized texts. For jaccard similarity, we represent each tokenized text as a set of tokens and then find the similarity using the formula in Eq. 4. For count-based approach, we create document vector based on the word counts of the words present in that document. To overcome the drawbacks in count-based approach, we use tf-idf method where documents are represented using the tf-idf scores (Eq. 1). For both these methods, the similarity is calculated using the cosine similarity (Eq. 5) and PCC (Eq. 6).

Word/Document Embedding: In order to capture the semantics, we move on to the embedding based techniques. For tf-idf+LSA methods, we first find the tf-idf scores to filter out unimportant words and then apply LSA to find the important topics where we choose from 20, 128 and 512 number of topics. For doc2vec methods, we choose different word embedding techniques like word2vec, GloVe and fasttext. Again, for all these methods, similarity is computed using cosine similarity and PCC.

String Matching: In this method, we do not tokenize the text unlike the above methods. We directly compare two text strings from *year1* and *year2* because we are trying to find the similarities between two texts as a human would find by matching two strings.

Purely Textual Features: For text only features, we follow four different methods. (1) *year1*: We consider the tokenized text from *year1* to predict the numeric score for *year2*. (2) *year2*: We follow similar approach as (1) with the only difference being tokenized text from *year2* to predict *year2* numeric scores. For both these methods, we represent the documents using tf-idf scores to create the document vectors. (3) *year1* + *year2* (concat): For each of the two text documents, we first create tf-idf vectors using the same approach as above for (1) and (2). We then concatenate them to create a longer vector. We use this textual representation to predict the *year2* numeric scores. (4) *year1* + *year2* (merge): Unlike (3), for two text documents, we first merge them as a regular text string. Then following the same approach for (1) and (2), we represent this merged text using a tf-idf vector which is then used to predict the *year2* scores.

Regression Results. We follow the methods described above to compute the similarities between two texts. The detailed experimental results are shown in Table 3. We do not report the PCC values in this table because we get identical results as cosine similarity in almost all the cases. Hence, we just report the mse for jaccard and cosine similarity values.

We can observe that using the delta values (i.e., *year2-year1* numeric values) performs the best overall on all variables except *Tobin's Q Ratio*. Here, we are using the difference in the values from one year to the next to predict next year's (*year2*) value, and so it is not entirely surprising that this is a strong baseline. It is also implicitly capturing information about the dependent variable (*year2* numeric score), and therefore is not entirely fair.

We find that among the textual approaches, the concatenation of *year1* and *year2* text performs the best for *ROA*, *EPS*, *Tobin's Q Ratio* and *Tier 1 Capital Ratio*. For *Tobin's Q Ratio*, we get better mse with purely textual features than the numeric score. For *Leverage Ratio* and *Z-score ratio*, we find that *year1* and *year2* textual features respectively, are the best among the textual approaches. We conclude that using textual features, especially concatenating the text from *year1* and *year2* is the overall best approach to predict the bank performance variables, and results in a better approach than using only the numeric scores.

Table 3. Regression based on financial text

	Bank variables/Methods	ROA	EPS	Tobin's Q ratio	Tier 1 capital ratio	Leverage ratio	Z-score
Baseline [numeric]	Year1	2.10E-03	2.89E-03	2.37E-03	**5.32E-04**	8.61E-04	1.03E-02
	Delta	**1.89E-03**	**2.64E-03**	4.27E-03	2.45E-03	**7.94E-04**	**7.39E-03**
Bag-of- Words	Jaccard similarity	2.60E-03	3.71E-03	4.78E-03	2.56E-03	1.11E-03	1.12E-02
	Count vectorizer	2.65E-03	3.78E-03	4.86E-03	2.58E-03	1.11E-03	1.14E-02
	tf-idf	2.66E-03	3.79E-03	4.87E-03	2.58E-03	1.11E-03	1.14E-02
Word/document embedding	tf-idf+ lsa(20)	2.66E-03	3.79E-03	4.87E-03	2.58E-03	1.11E-03	1.14E-02
	tf-idf+ lsa(128)	2.66E-03	3.79E-03	4.87E-03	2.58E-03	1.11E-03	1.14E-02
	tf-idf+ lsa(512)	2.66E-03	3.79E-03	4.87E-03	2.58E-03	1.11E-03	1.14E-02
	doc2vec (word2vec)	2.62E-03	3.73E-03	4.79E-03	2.57E-03	1.11E-03	1.13E-02
	doc2vec (glove)	2.62E-03	3.73E-03	4.79E-03	2.57E-03	1.11E-03	1.13E-02
	doc2vec (fasttext)	2.62E-03	3.73E-03	4.78E-03	2.57E-03	1.11E-03	1.12E-02
String matching	Fuzzy- wuzzy	2.65E-03	3.77E-03	4.88E-03	2.58E-03	1.11E-03	1.15E-02
Year1	tf-idf	2.13E-03	2.91E-03	2.22E-03	8.71E-04	8.79E-04	9.61E-03
Year2	tf-idf	2.10E-03	2.70E-03	2.22E-03	7.77E-04	9.02E-04	9.38E-03
Year1 + year2 (concat)	tf-idf	1.93E-03	2.66E-03	**1.92E-03**	6.38E-04	8.91E-04	9.70E-03
Year1 + year2 (merge)	tf-idf	2.11E-03	2.78E-03	2.22E-03	7.86E-04	9.34E-04	9.48E-03

5 Conclusion and Future Work

We performed a comparative study of different textual similarity techniques to compare long text documents such as Sects. 7 and 7A of 10-K filings from two consecutive years. We divided these techniques into simple bag-of-words, word/document embeddings and string matching types. Since the bag-of-words model does not capture any semantics of the text, we follow the word and document embedding approach to create document vectors. From our comparative experiments, we observed that doc2vec (using both, cosine similarity and PCC) and jaccard similarity show similar trends. We also saw a slight drop in similarity for years 2008–2010 and we can therefore say that the documents between the years 2008 and 2010 were not as similar when compared to later years. Further, we used these similarity scores to predict the next year's values of different bank variables to see if the textual changes by themselves are predictive. We observe that the combined text from the previous and current years is helpful in predicting several bank variables, namely *ROA*, *EPS*, *Tobin's Q Ratio* and *Z-score*. In future, we plan to use more advanced document embedding methods such as temporal topic models to better understand the similarities qualitatively between the topics present in a pair of documents. It would be also interesting to extend the time period beyond consecutive year pairs to a longer time window. Since the document length increases over the years, using normalized document length can also be used as an additional feature.

Acknowledgments. This work was supported in part by NSF Award III-1738895.

References

1. Agirre, E., Cer, D., Diab, M., Gonzalez-Agirre, A.: Semeval-2012 task 6: a pilot on semantic textual similarity. In: * SEM 2012: The First Joint Conference on Lexical and Computational Semantics-Volume 1: Proceedings of the main conference and the shared task, and Volume 2: Proceedings of the Sixth International Workshop on Semantic Evaluation (SemEval 2012), pp. 385–393 (2012)
2. Alschner, W.: Sense and similarity: automating legal text comparison. Available at SSRN 3338718 (2019)
3. Benedetti, F., Beneventano, D., Bergamaschi, S., Simonini, G.: Computing inter-document similarity with context semantic analysis. Inf. Syst. **80**, 136–147 (2019)
4. Blei, D.M., Lafferty, J.D.: Dynamic topic models. In: Proceedings of the 23rd international conference on Machine learning, pp. 113–120 (2006)
5. Brown, S.V., Ma, G., Tucker, J.W.: A measure of financial statement similarity. Available at SSRN 3384394 (2019)
6. Brown, S.V., Tucker, J.W.: Large-sample evidence on firms' Year-over-year MD&A modifications. J. Acc. Res. **49**(2), 309–346 (2011)
7. Deerwester, S., Dumais, S.T., Furnas, G.W., Landauer, T.K., Harshman, R.: Indexing by latent semantic analysis. J. Am. Soc. Inf. Sci. **41**(6), 391–407 (1990)
8. Dyer, T., Lang, M., Stice-Lawrence, L.: The evolution of 10-k textual disclosure: evidence from latent Dirichlet allocation. J. Account. Econ. **64**(2–3), 221–245 (2017)

9. Gong, H., Sakakini, T., Bhat, S., Xiong, J.: Document similarity for texts of varying lengths via hidden topics. arXiv preprint arXiv:1903.10675 (2019)

10. Gupta, A., Owusu, A.: Identifying the risk culture of banks using machine learning. Available at SSRN 3441861 (2019)

11. He, Y., Li, J., Song, Y., He, M., Peng, H., et al.: Time-evolving text classification with deep neural networks. In: IJCAI, pp. 2241–2247 (2018)

12. Huang, X., Paul, M.: Examining temporality in document classification. In: Proceedings of the 56th Annual Meeting of the Association for Computational Linguistics, vol. 2, pp. 694–699 (2018)

13. Hurst, M.F., et al.: Temporal text mining. In: AAAI Spring Symposium: Computational Approaches to Analyzing Weblogs, pp. 73–77 (2006)

14. Joulin, A., Grave, E., Bojanowski, P., Mikolov, T.: Bag of tricks for efficient text classification. arXiv preprint arXiv:1607.01759 (2016)

15. Kenter, T., De Rijke, M.: Short text similarity with word embeddings. In: Proceedings of the 24th ACM International on Conference on Information and Knowledge Management, pp. 1411–1420 (2015)

16. Kogan, S., Levin, D., Routledge, B.R., Sagi, J.S., Smith, N.A.: Predicting risk from financial reports with regression. In: Proceedings of Human Language Technologies: The 2009 Annual Conference of the North American Chapter of the Association for Computational Linguistics, pp. 272–280 (2009)

17. Kusner, M., Sun, Y., Kolkin, N., Weinberger, K.: From word embeddings to document distances. In: International Conference on Machine Learning, pp. 957–966 (2015)

18. Le, Q., Mikolov, T.: Distributed representations of sentences and documents. In: International Conference on Machine Learning, pp. 1188–1196 (2014)

19. Li, H.: Repetitive disclosures in the MD&A. J. Bus. Finan. Acc. **46**(9–10), 1063–1096 (2019)

20. Liu, M., Lang, B., Gu, Z.: Calculating semantic similarity between academic articles using topic event and ontology. arXiv preprint arXiv:1711.11508 (2017)

21. Liu, M., Lang, B., Gu, Z., Zeeshan, A.: Measuring similarity of academic articles with semantic profile and joint word embedding. Tsinghua Sci. Technol. **22**(6), 619–632 (2017)

22. Mandal, A., Chaki, R., Saha, S., Ghosh, K., Pal, A., Ghosh, S.: Measuring similarity among legal court case documents. In: Proceedings of the 10th Annual ACM India Compute Conference, pp. 1–9 (2017)

23. Mikolov, T., Chen, K., Corrado, G., Dean, J.: Efficient estimation of word representations in vector space. arXiv preprint arXiv:1301.3781 (2013)

24. Momeni, E., Karunasekera, S., Goyal, P., Lerman, K.: Modeling evolution of topics in large-scale temporal text corpora. In: Twelfth International AAAI Conference on Web and Social Media (2018)

25. Müller, M.C.: Semantic matching of documents from heterogeneous collections: a simple and transparent method for practical applications. arXiv preprint arXiv:1904.12550 (2019)

26. Nguyen, H.T., Duong, P.H., Cambria, E.: Learning short-text semantic similarity with word embeddings and external knowledge sources. Knowl.-Based Syst. **182**, 104842 (2019)

27. Oghbaie, M., Mohammadi Zanjireh, M.: Pairwise document similarity measure based on present term set. J. Big Data **5**(1), 1–23 (2018). https://doi.org/10.1186/s40537-018-0163-2

28. Pennington, J., Socher, R., Manning, C.D.: GloVe: global vectors for word representation. In: Proceedings of the 2014 Conference on Empirical Methods in Natural Language Processing (EMNLP), pp. 1532–1543 (2014)
29. Pontes, E.L., Huet, S., Linhares, A.C., Torres-Moreno, J.M.: Predicting the semantic textual similarity with Siamese CNN and LSTM. arXiv preprint arXiv:1810.10641 (2018)
30. Rawte, V., Gupta, A., Zaki, M.J.: Analysis of year-over-year changes in risk factors disclosure in 10-k filings. In: Proceedings of the Fourth International Workshop on Data Science for Macro-Modeling with Financial and Economic Datasets, pp. 1–4 (2018)
31. Renjit, S., Idicula, S.M.: Cusat nlp@ aila-fire2019: similarity in legal texts using document level embeddings. In: Proceedings of FIRE (2019)
32. Shahmirzadi, O., Lugowski, A., Younge, K.: Text similarity in vector space models: a comparative study. In: 2019 18th IEEE International Conference On Machine Learning And Applications (ICMLA), pp. 659–666. IEEE (2019)
33. Sugathadasa, K., et al.: Legal document retrieval using document vector embeddings and deep learning. In: Arai, K., Kapoor, S., Bhatia, R. (eds.) SAI 2018. AISC, vol. 857, pp. 160–175. Springer, Cham (2019). https://doi.org/10.1007/978-3-030-01177-2_12
34. Vashishth, S., Dasgupta, S.S., Ray, S.N., Talukdar, P.: Dating documents using graph convolution networks. arXiv preprint arXiv:1902.00175 (2019)
35. Vega, D., Magnani, M.: Foundations of temporal text networks. Appl. Netw. Sci. **3**(1), 25 (2018)
36. Wang, Y., et al.: MedSTS: a resource for clinical semantic textual similarity. Lang. Resour. Eval. 1–16 (2018)
37. Wright, R.: Temporal Text Mining: A Thematic Exploration of Don Quixote (2017)
38. Zheng, T., et al.: Detection of medical text semantic similarity based on convolutional neural network. BMC Med. Inform. Decis. Mak. **19**(1), 156 (2019)

Ranking Cryptocurrencies by Brand Importance: A Social Media Analysis in ENEAGRID

Giuseppe Santomauro[1]([✉]), Daniela Alderuccio[2]([✉]), Fiorenzo Ambrosino[1]([✉]), and Silvio Migliori[3]([✉])

[1] ENEA - C.R. Portici, DTE-ICT-HPC, P.le E. Fermi, 1, 80055 Portici, NA, Italy
{giuseppe.santomauro,fiorenzo.ambrosino}@enea.it
[2] ENEA - Sede Legale, DTE-ICT-HPC, L. Thaon di Revel, 76, 00196 Rome, Italy
daniela.alderuccio@enea.it
[3] ENEA - Sede Legale, DTE-ICT, L. Thaon di Revel, 76, 00196 Rome, Italy
silvio.migliori@enea.it

Abstract. This paper presents a preliminary analysis of cryptocurrency brands on Twitter, carried out through the Semantic Brand Score Brand Intelligence App (a tool hosted in the ENEAGRID digital infrastructure). The aim is to rank five digital coins (i.e. Bitcoin, Ethereum, Zcash, Monero, Litecoin). Web crawling, data storage and brand scoring activities require computational power. The ENEAGRID infrastructure faces this challenge in terms of computational costs, with its computing core represented by the HPC CRESCO clusters. In our methodology, we run periodic sessions of web crawling to create a database of tweets (concerning digital coins), then we use the Semantic Brand Score to evaluate cryptocurrencies relevance and study their brand image. This case study is a first step towards collaborating with experts and research communities in financial domain and opening access to ENEA Virtual Labs.

Keywords: Social network analysis · Semantic brand score · Cryptocurrency · Financial text mining

1 Introduction

In this exploratory analysis, we present the results of ranking digital currencies using Twitter data. We use the Semantic Brand Score Brand Intelligence (SBS BI) app (Fronzetti Colladon and Grippa 2020) to assign a brand importance score (Fronzetti Colladon 2018) to five different digital coins. The ENEA web crawling tool has been extended to retrieve data from social media, in order to mine information in the financial domain. ENEA computational and storage power is used to crawl and store data in the ENEAGRID digital infrastructure (hosting the SBS BI app to enhance its computational power). Brand scoring is the task of performing a brand ranking from a large set of textual data. By capturing signals from tweets, we could assess the brand importance of Bitcoin, Zcash, Ethereum, Litecoin and Monero, considering: the frequency of use of

© Springer Nature Switzerland AG 2021
V. Bitetta et al. (Eds.): MIDAS 2020, LNAI 12591, pp. 92–100, 2021.
https://doi.org/10.1007/978-3-030-66981-2_8

their brand name; the variety of the words associated with the brands; the brands ability to bridge connections between other words or topics. This paper presents a preliminary experiment carried out considering tweets produced in the first week of the month, from January to June 2019.

1.1 Cryptocurrencies

Cryptocurrencies are electronic cash systems administrated via peer-to-peer consensus, without a central authority. They are supported by a decentralized mechanism, which is independent of governmental functions and available anywhere in the world. Due to lack of sufficient regulation, cryptocurrencies may be both tightly related to economic activity and subject to manipulations by sophisticated investors. Cryptocurrencies use High Performance Computing to verify transactions (mining), convert them into blocks, and add them to the blockchain. Mining is based on cryptographic algorithm assuring anonymity and payment untraceability. Crypto-market is emotional and irrational (e.g. the Giffen paradox); people may respond to the threat of global instability by switching from traditional currencies to cryptocurrencies. Cryptocurrencies volatility levels are usually much higher than traditional currencies. It is recognized that there is a link between the economic policy uncertainty index (EPU) (Baker et al. 2016) and cryptocurrency volatility (Yen and Cheng 2020). Investment results can be reached through diversification portfolio, that is setting a portfolio of different cryptocurrencies, to be re-balanced when advantageous. Matta et al. (2015) found a correlation link between Google searches related to Bitcoin and prices, so that Google Trends could be seen as a kind of predictor. They used Google Trends media to analyse Bitcoin's popularity under the perspective of Web search and found significant cross correlation values, especially between Bitcoin price and Google Trends data.

In this paper we evaluate cryptocurrencies brand score (that is their brand importance). Then we pose the question whether measuring a cryptocurrency brand score can help in finding information on cryptocurrencies dominance in the crypto-market. After that, we try to find correlation among digital coins, in order to re-balance a portfolio. Cryptocurrencies are highly connected to each other but not connected to other financial assets and - in order to reduce investment risks - several cryptocurrencies might be needed as a set, by adding several different crypto-currencies to a portfolio.

1.2 Twitter as a Financial Source of Information

Social Media allows the sharing of knowledge or information about a topic or an event. In Social Media Platform thoughts and ideas find a sort of collective arena, where people connect and share idea. For this reason, since the beginning the crypto-industry has thrived on them. Twitter is an online social networking website with an informative function, attracting the increasing interest of individual user to share information on investigating decisions. Steinert and Herff (2018) suggests a relationship between social media and cryptocurrencies prices: altcoin returns can be predicted to some extent by using the information provided by Twitter. The aim of their models was to analyse whether it is possible to predict altcoin returns based upon social media activity. The activity was measured by the number of tweets referring to a certain altcoin and the sentiment. In this exploratory analysis we take a snapshot of what the world is talking about cryptocurrencies, seen through the eyes of Twitter.

2   Text Mining and Analytics in ENEAGRID

Web (and Social Network) crawling and the SBS BI app require significant computational power. The ENEAGRID/CRESCO is a digital infrastructure, providing: computing power, data storage and it is based on distributed resources in six Data Centers (Migliori et al. 2019), interconnected via GARR network. ENEAGRID offers the "Text Mining & Analytics Platform", an integrated platform of high-profile technologies and e-tools for crawling, storing, indexing, analyzing, extracting, visualizing data. Its analytical power is complemented by the SBS BI app, a tool primarily accessible at http:// bi.semanticbrandscore.com and now also hosted on ENEAGRID[1]. While SBS BI is fully described in the work of Fronzetti Colladon and Grippa (2020), below we describe the "Web Crawling" virtual lab, which was used to collect data subsequently analyzed through the SBS BI app. We created a collaborative "Web Crawling" Project integrated in the Text Mining & Analytics Platform of ENEAGRID. Generally, Web crawling is the activity to browse www systematically, download web content and store crawled data. In the "Web Crawling" Virtual Laboratory (Santomauro et al. 2020) is described the adoption of BUbiNG, an open source software that allows the parallelization of multiple crawling agents (Boldi et al. 2016). Recently, we extended our V-Lab with an instrument for the Social Network Crawling, in this case, specifically for Twitter, based on the free APIs of the Social Network.

3   The SBS Index

The Semantic Brand Score (SBS) measures brand importance, which is at the basis of brand equity. SBS metric was partially inspired by conceptualizations of brand equity and by the constructs of brand image and brand awareness (Aaker 1996; Keller 1993). Brand importance is measured along three dimensions: prevalence, diversity and connectivity (Fronzetti Colladon 2018). Prevalence measures the number of times (frequency) a brand is directly mentioned. Diversity measures the richness of its lexical embeddings. Connectivity represents the brand ability to bridge connections between other words or discourse topics. Brand importance, as measured by the Semantic Brand Score, has already proven its usefulness in different contexts, such as predicting electoral outcomes (Fronzetti Colladon 2020) or visits to touristic attractions (Fronzetti Colladon et al. 2020). The SBS metric is fully described in the work of Fronzetti Colladon (2018), with a variation suggested in a later work, for the calculation of diversity. Accordingly, in this paper we use the distinctiveness centrality metric (Fronzetti Colladon and Naldi 2020), in lieu of degree centrality (Freeman 1979) to calculate diversity. Distinctiveness not only captures the heterogeneity of textual brand associations, but also the uniqueness of their strength.

---

[1] http://enea.semanticbrandscore.com.

4 Case Study: Ranking Cryptocurrencies on Twitter

ENEAGRID is equipped with a social media crawler that downloads contents from Twitter and hosts the SBS BI app, which uses the ENEA computational and storage power. The SBS metric applied to digital coins ranks Cryptocurrencies from Twitter data, and assess the distribution of tweets of five cryptocurrencies (Bitcoin, Ethereum, Zcash, Monero, Litecoin). For each digital currency, we studied: how many tweets per digital coin; in which context, different cryptocurrencies were mentioned; how each digital coin was connected with other ones (useful for re-balancing the portfolio diversification, so lowering investment risks).

4.1 Social Media Analysis in ENEAGRID

In Social Media space people interact to express their opinion and share experience; users play many roles and define significant relationships, also useful in Economy, Finance, Marketing, Political Campaign, etc. Some interesting research questions are: (i) by measuring brand ranking in a certain moment or in a time frame, do we have information on the dominance of a single digital coin in the market? (ii) is the value of a Cryptocurrency related to the volumes (number) of its tweets?

Social Network Crawling in ENEAGRID
In ENEAGRID we opened a new virtual environment on Twitter data to rank digital currencies. We planned running periodic sessions of crawling, the first week of each month from January 2019 to June 2019. So we created a database of tweets, to be analysed by the Semantic Brand Score. Then we started creating dataset on a specific topic (cryptocurrencies) and downloading and storing texts from Social Networks (in this case study: from Twitter). Our Methodology steps were: running periodic sessions of social crawling in order to create a database of tweets that concern news and discussions about digital coins; use of parallel developer accounts to avoid the limitation on the number of tweets downloaded per user; applying the Semantic Brand Score to rank cryptocurrency importance; creating a rank among five most popular cryptocurrencies (i.e. Bitcoin, Ethereum, Zcash, Monero, Litecoin). Data crawled amount is 333K tweets in total (average: 8K tweets/day). We observed that, for every day considered for retrieving data, due the limits of free APIs, our crawler downloaded in backwards tweets posted between 23:59 and around 18:30 (an average of 5.5 h on 24 h at day). By supposing an uniform distribution of tweets posted along a day, we estimated we collected an average of 23% (5,5 h/24 h) with respect to total amount of tweets for day. Data were stored in a specific dedicated area in ENEAGRID. Data output was set to .csv for easy data manipulation and compliant with the SBS input format. Each line of csv file is a composed by three separated fields (generally the separator is the pipe character "|"): the text of a tweet; the date of its publication; the weight of tweet (cfr. the following first 2 lines of .csv file of our experiment):

"$BTC Over 900 Retailers Now Accepting Bitcoin Cash http://twib.in/l/rMGXMykGRej5" | 01/01/2019 | 1
"Price of bitcoin must tens to zero as there is no economic activity underpinning it" | 01/01/2019 | 1

Semantic brand score is independent on quantity because it represents the importance of a digital coin in terms of brand score – and its values are standardized to allow the comparison of different datasets.

4.2 SBS Index Application for Measuring Cryptocurrency Brand Scoring

Applying the Semantic Brand Score to rank cryptocurrency importance of five digital coins, the SBS report shows: SBS Time Trends, that is the SBS evolution over time, also with respect to other brands (Fig. 1); Brand Positioning; the contribution to brand importance of the three SBS dimensions (Prevalence, Diversity and Connectivity) (Fig. 2); most frequent words for each time period, most frequent brand associations; Trend of unique brand associations, extending the concept of Diversity (this trend dynamic will be very interesting for the detection of new emergent topics, possibly with a forecasting power); Brand Image Similarity; Target words for the improvement of connectivity; the Network of the main discourse topics; Topic relevance (Fig. 3) and connections among topics (showing how strong are the connections among topic); Brand and Topics prevalence (Fig. 4).

4.3 Results

In this exploratory analysis, by measuring brand importance we expect to find information on cryptocurrencies dominance in the crypto-market, and trends over six months. For this reason, we search signals from Social media (i.e. Twitter), analysing tweets. Data were collected from Twitter over prolonged time frame, set to one week for a period of 6 months, ranging from January to June 2019. Tag words for search query are as follows: "cryptocurrency OR cryptocurrencynews OR CryptoNews OR bitcoin OR Zcash OR Ethereum OR Litecoin OR Monero". We pre-process tweets and remove space, stopwords, negations, punctuation, symbols (#, @), and anonymize Tweets IDs. Mentions and hashtags aren't evidenced by symbols, but their importance is highlighted in the network. In this first exploratory analysis, we do not scrape Twitter data from verified account and we do not take into consideration user sentiment. We will try to give

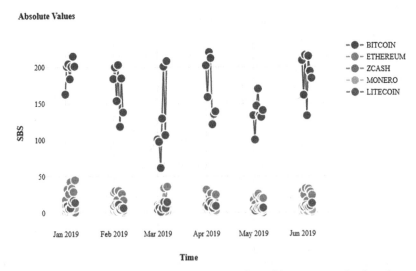

Fig. 1. SBS time trends - SBS evolution over time with respect to other brands.

an interpretation to some variations in trends in certain moment, searching for political or socio-economic events.

Data Analysis, Visualization and Interpretation
In this case-study we take into consideration five digital coin: Bitcoin, Ethereum, Zcash, Monero, Litecoin, crawling Twitter data of the first week of each of the six months. In Fig. 1 you will find the SBS evolution over six months. It shows the dominance of Bitcoin on the crypto-market.

Figure 2 shows the contribution of the three dimensions of Prevalence, Diversity and Connectivity to the Brand Importance. Bitcoin shows a balanced distribution of SBS dimensions.

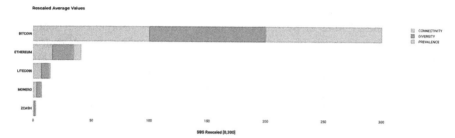

Fig. 2. Brand importance: contribution of prevalence, diversity and connectivity.

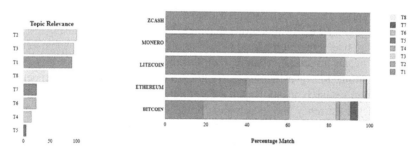

Fig. 3. Topic relevance in [0,100].

Fig. 4. Brands and Topics in [0,100].

From the set of words representative of each topic generated by SBS, we can get some insights on the exposure of the considered cryptocurrencies to those topics. For example: Topic 1 (T1) include words related to the price of the cryptocurrencies as it includes also the price tickers; Topic 2 (T2) is more oriented to payment capabilities other than market trends; Topic 3 (T3) is more oriented to platform functionalities like tokens generation and smart contracts and so on. For the sake of brevity, we omit the description of topics from 4 to 8 that are less relevant. We can see that Bitcoin is the cryptocurrency that covers in a more balanced way all the emerged topics; it is also the only one that is not prevalent on the T1 topic. T2 is the more important for Bitcoin as it is the first cryptocurrency and its main objective is to be a payment network. We can see also the relevance of T3 for Ethereum as it has been the most known platform for smart contract functionalities.

5 Future Steps

We will extend this exploratory analysis to the first semester 2020 and we expect to find signals of relevant events, having strong socio-economic effects in 2020, such as: the halving of Bitcoin value happening every four years, and the sanitary emergence caused by COVID-19. We plan to extend the crawling, so capturing the effect of the Halvening of mid-May 2020, and the re-opening of European borders after the lockdown, expecting to find how these events impact in tweets. Both researches will be published in separate studies. Further steps of this research are towards: applying user sentiment (by implementing a specific financial dictionary); leveraging the power of distributed computing, and deep learning (Mariano et al. 2018).

6 Discussion and Conclusion

In this case study on cryptocurrencies, the ENEA web crawling tool has been extended to retrieve data from social media, in order to discover information in financial domain. ENEA computational and storage power is used to crawl and store data. We complemented our framework with the SBS BI app, a tool to calculate the Semantic Brand Score and other brand analytics (Fronzetti Colladon and Grippa 2020). The Semantic Brand Score metric applied to cryptocurrencies tries to evaluate the importance trends of digital coins in the cryptomarket. Results shows that ranking cryptocurrencies using the Semantic Brand Score over a medium period looks promising in finding how socio-economic and political events affect the cryptomarket. Starting from the results of this research, we plan to continue our investigation in cryptocurrencies domain and we expect to find answers to other research question such as: (i) finding a correlation between number of tweets of a cryptocurrency and its semantic brand score; (ii) finding whether it is possible to forecast a cryptocurrency value by measuring its brand ranking dynamic. Due to the fact that cryptocurrencies are subject to economic and political uncertainty, the investigation on digital coins over a longer period of time helps in finding which events affect variations in cryptocurrency trends, finding correlation among socio-economic indicators. For this reason, in the future by tracing dynamic trends over 12–18 months, we expect to find elements useful to forecast trends and impacts of future events in cryptocurrencies, and correlation between tweets volume variations and price trends variations of digital coins (Bollen et al. (2011) demonstrated that tweet sentiment can predict the market trend 3–4 days in advance, with a good chance of success). With this research activity ENEA opens to collaborations with Research Communities interested in Financial Mining of Massive Volume of Text Data (PB &TB-sized corpora) in Economic-Social Sciences, Big Data, Web Mining, etc.

Acknowledgements. The authors are grateful to Andrea Fronzetti Colladon, professor at the University of Perugia, who has been collaborating with their team at ENEA for more than two years and who provided technical and theoretical support for the writing of this paper. The computing resources and the related technical support used for this work have been provided by CRESCO/ENEAGRID High Performance Computing infrastructure and its staff Ponti et al. (2014). CRESCO/ENEAGRID HPC infrastructure is funded by ENEA, the Italian National Agency for New Technologies, Energy and Sustainable Economic Development and by Italian and European research programmes, see http://www.cresco.enea.it/english for information.

References

Aaker, D.A.: Measuring brand equity across products and markets. Calif. Manag. Rev. **38**(3), 102–120 (1996)

Baker, S.R., Bloom, N., Davis, S.J.: Measuring economic policy uncertainty. Q. J. Econ. **131**(4), 1593–1636 (2016). https://doi.org/10.1093/qje/qjw024. https://academic.oup.com/qje/article/131/4/1593/2468873

Boldi, P., et al.: BUbiNG: Massive Crawling for the Masses (2016)

Bollen, J., Mao, H., Zeng, X.: Twitter mood predicts the stock market. J. Comput. Sci. **2**(1), 1–8 (2011). https://doi.org/10.1016/j.jocs.2010.12.007

Freeman, L.C.: Centrality in social networks conceptual clarification. Soc. Netw. **1**(3), 215–239 (1979). https://doi.org/10.1016/0378-8733(78)90021-7

Fronzetti Colladon, A., Grippa, F.: Brand Intelligence analytics. In: Przegalinska, A., Grippa, F., Gloor, P.A. (eds.) COINs 2019. SPC, pp. 125–141. Springer, Cham (2020). https://doi.org/10.1007/978-3-030-48993-9_10

Fronzetti Colladon, A.: The semantic brand score. J. Bus. Res. **88**, 150–160 (2018). https://doi.org/10.1016/j.jbusres.2018.03.026

Fronzetti Colladon, A.: Forecasting election results by studying brand importance in online news. Int. J. Forecast. **36**(2), 414–427 (2020)

Colladon, A.F., Grippa, F., Innarella, R.: Studying the association of online brand importance with museum visitors: an application of the semantic brand score. Tour. Manag. Perspect. **33**, 100588 (2020). https://doi.org/10.1016/j.tmp.2019.100588

Fronzetti Colladon, A., Naldi, M.: Distinctiveness centrality in social networks. PLOS ONE **15**(5), e0233276 (2020). https://doi.org/10.1371/journal.pone.0233276

Keller, K.L.: Conceptualizing, measuring, and managing customer-based brand equity. J. Mark. **57**(1), 1–22 (1993)

Yen, K.-C., Cheng, H.-P.: Economic policy uncertainty and cryptocurrency volatility. https://www.sciencedirect.com/science/article/abs/pii/S1544612319310189

Matta, M., Lunesu, I., Marchesi, M.: Bitcoin spread prediction using social and web search media. In: Conference: Workshop Deep Content Analytics Techniques for Personalized and Intelligent Services (2015)

Mariano, A., Adani, M., Briganti, G., Mircea, M.: Qualità dell'aria: Algoritmi di Machine Learning applicati a dati simulati dal sistema modellistico AMS-MINNI. In: Conference on GARR 'Data Revolution', pp. 75–81 (2018). https://doi.org/10.26314/garr-conf2018-proceedings-14

Migliori, S., et al.: CRESCO HPC Clusters Evolution, Users and Workloads 2008–2018 - in High Performance Computing on CRESCO Infrastructure: Research Activity Results 2018 (2019). https://ict.enea.it/wp-content/uploads/2020/02/report_cresco_2018_for_web.pdf. ISBN: 978-88-8286-390-6

Ponti, G., et al.: The role of medium size facilities in the HPC ecosystem: the case of the new CRESCO4 cluster integrated in the ENEAGRID infrastructure. In: Proceedings of the 2014 International Conference on High Performance Computing and Simulation, HPCS 2014, Art. no. 6903807, pp. 1030–1033 (2014)

Santomauro, G., Alderuccio, D., Ambrosino, F., Fronzetti Colladon, A., Migliori, S.: A brand scoring system for cryptocurrencies based on social media data. In: Bitetta, V., Bordino, I., Ferretti, A., Gullo, F., Pascolutti, S., Ponti, G. (eds.) MIDAS 2019. LNCS (LNAI), vol. 11985, pp. 127–132. Springer, Cham (2020). https://doi.org/10.1007/978-3-030-37720-5_11

Steinert, L., Herff, C.: Predicting altcoin returns using social media. PLoS ONE **13**(12), e0208119 (2018). https://doi.org/10.1371/journal.pone.0208119

Towards the Prediction of Electricity Prices at the Intraday Market Using Shallow and Deep-Learning Methods

Christoph Scholz$^{(\boxtimes)}$ ⓘ, Malte Lehna ⓘ, Katharina Brauns, and André Baier

Fraunhofer IEE, 34119 Kassel, Germany
{christoph.scholz,malte.lehna,katharina.brauns,
andre.baier}@iee.fraunhofer.de

Abstract. The percentage of renewable energies (RE) within power generation in Germany has increased significantly since 2010 from 16.6% to 42.9% in 2019 which led to a larger variability in the electricity prices. In particular, generation from wind and photovoltaics induces high volatility, is difficult to forecast and challenging to plan. To counter this variability, the continuous intraday market at the *EPEX SPOT* offers the possibility to trade energy in a short-term perspective, and enables the adjustment of earlier trading errors. In this context, appropriate price forecasts are important to improve the trading decisions on the energy market. Therefore, we present and analyse in this paper a novel approach for the prediction of the energy price for the continuous intraday market at the *EPEX SPOT*. To model the continuous intraday price, we introduce a semi-continuous framework based on a rolling window approach. For the prediction task we utilise shallow learning techniques and present a *LSTM*-based deep learning architecture. All approaches are compared against two baseline methods which are simply current intraday prices at different aggregation levels. We show that our novel approaches significantly outperform the considered baseline models. In addition to the general results, we further present an extension in form of a multi-step ahead forecast.

1 Introduction

Since the liberalization of the European electricity markets in the 1990s, the primary trading location for electricity commodities are the national day-ahead spot markets. However, with the increased electricity production through renewable energies, market prices have become more volatile. As result a rising demand for continuous trading on intraday spot markets emerged in recent years.[1] On the intraday spot markets, participants are able to account for rapid changes

---

[1] http://static.epexspot.com/document/38579/Epex_TradingBrochure_180129_Web.pdf.

C. Scholz and M. Lehna—Both authors contributed equally to this work.

© Springer Nature Switzerland AG 2021
V. Bitetta et al. (Eds.): MIDAS 2020, LNAI 12591, pp. 101–118, 2021.
https://doi.org/10.1007/978-3-030-66981-2_9

in demand and supply and integrate these changes in the pricing on a short-term notice. However, in current research the continuous intraday market is still under-represented, especially in the electricity price forecast (EPF). Even though various research has been conducted on different day-ahead markets, as reviewed by Gürtler and Paulsen [6], to the best of our knowledge only a small amount of papers [1,4,9,13,14,18] have been published for the price forecast on the intraday market. Furthermore, all these papers only focus on forecasting the final price on the continuous intraday market. Consequently, this paper is intended to contribute new insights for EPF into the intraday market. We propose a novel (semi-continuous) forecast framework in order to capture a more consecutive representation of the intraday market. In addition, we incorporate previous research proposals by combining several external factors in our forecast study. Finally, different machine learning methods are presented and analyzed regarding their predictive power.

1.1   Research Environment

As research environment, we chose the intraday spot market of Germany from the year 2018, as we had enough data available to investigate the continuous nature of the spot price and realize the respective forecast. On the market, the electricity for the same day delivery is traded in two different products, which are the one-hour and quarterly-hour products, and are based on their delivery length. For each of the one-hour (or quarterly-hour) products of the day, the market participants trade in different magnitudes of electricity feed-in through an continuous orderbook system. However, in the continuous intraday market the fundamental challenge of the EPF is the short time horizon in which the products are traded. Between the opening of the market and the termination by the delivery, only a small time interval is available for trading. In terms of hourly intervals, the market opening occurs on the previous day at 15:00 while the trade for the quarterly intervals begins at 16:00. For both, the products are traded continuously up until 30 min before the delivery on the whole market and up to 5 min within the control areas of the distributors. In addition, the majority of transactions take place in the last hours before delivery. In Fig. 1 this effect is visualized, where one can clearly see the increase in trading volume over time. Moreover, note the two drops in the graph which correspond to the described 30 min and 5 min mark. Due the fact that the trading volume gets stronger towards the end, we focus in our scenario on the last four hours till delivery time.

1.2   Contribution

Framework: Given the specific structure of the intraday market, we introduce a semi-continuous framework to enable the forecasting of the continuous intraday prices. However, instead of only focusing on one value of each product (as done in previous papers like [9,14,18] for the German market), we propose to forecast the electricity price development at regular intervals. This allows to forecast multiple observations per product and give a more precise description of the price development. Our intention behind this short-term perspective is the ability to forecast the price movement, while the product is still traded on the market. In comparison to other EPF literature this approach is unique, due to the fact that until now researchers only predicted one value per product.

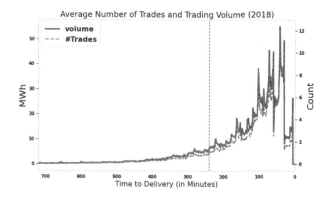

Fig. 1. Average number of both trades and traded volume for time t. The x-axis denotes the remaining time t (in minutes) till the delivery of the product. The left y-axis represent the average number of traded volume at time t, the right y-axis the average number of trades at time t. The red dotted line at minute 240 is the starting time of our forecast scenario.

Forecasting Approaches: One major consequence following from the structure of the research environment is that the implementation of time series approaches such as ARIMA and VAR models proves to be difficult. Due to their reliance on previous observations, it would be necessary to determine the auto-regressive order structure and provide sufficient amount of data to estimate the models for each individual product. As a result, in this paper we disregard the usage of basic time series model and instead analyze three different machine learning techniques for the EPF task. As model candidates, we decided to implement two shallow learning models as well as one deep learning model for the comparison analysis. In terms of the shallow learning models, the *Random Forests* [2] as a simple model and the *XG-Boost* [3] as a more advanced model were selected. However, it is important to note that while both approaches can be applied on a regression setting, they are not designed to capture time related dependencies between the samples i.e. auto-correlation or auto-regressive structures. Nevertheless, due to their flexibility and their robust behavior in terms of noise variables, we employ the two approaches. On the other hand, we decided to employ a *LSTM*-based deep neural network architecture for the prediction of the spot price. This network is well known to detect and use time dependencies for the prediction task. All models are further discussed in Sect. 4.2.

Outline: In general, our contribution in this paper can be summarized as follows:

1. We present the first work forecasting the electricity price development of individual products on the continuous intraday market.
2. We utilize three different machine learning approaches, two shallow learning and one *LSTM*-based deep learning approach and compare the predictive power with state-of-the-art baseline models.
3. We show that our presented novel approaches significantly outperform baseline models by at least 10%.
4. We perform and discuss the performance of a multi-step ahead forecast.

2   Related Work

Within the literature, intraday spot price receives more and more attention in recent years. Kiesel and Paraschiv [10] examine the biding behaviour of participants on the German intraday market based on 15 min products and analyze the different influencing factors. In their paper, they especially emphasise the importance of wind and photovoltaic forecast errors as one major impact for the trading behaviour. Similar results were published by Ziel [21], who analyzed the effect of wind and solar forecasting errors. Further analyses on the German intraday market were published [8,16], however with a different research focus. While Pape et al. [16] focused their research on the fundamentals of the electricity production and their influence on the intraday spot price, Kath [8] analyzed the effect of cross-border trade of electricity. Next to the german intraday market, other research was conducted on the Iberian electricity market [1,4,13] as well as the Turkish intraday market [15]. On the Iberian spot market MIBEL, the first advances in terms of intraday EPF have been made. While Monteiro et al. [13] use a MLP neural network to forecast the intraday prices, Andrade et al. [1] conduct a probabilistic forecasting approach to further analyze the influence of external data. Further research was published by Oksuz and Ugurlu[15], who examined the performance of different neural networks in the EPF in comparison to regression and LASSO techniques on the Turkish intraday market. In terms of the German, the intraday trade was primarily covered by three papers [9,14,18]. Kath and Ziel [9] predicted both the intraday continuous and the intraday call-auction prices of German quarterly hour deliveries based on an elastic net regression. Furthermore, they proposed a trading strategy based on their forecast results to realize possible profits. A different approach was proposed by Uniejewski et al. [18] which analyzed the German intraday spot price in form of the closure price of the *EPEX SPOT* ID3 index. For their forecast they employed a multivariate elastic net regression model to further conduct a variable selection and proposed a simple trading strategy to realize possible gains. With a very similar research setting Narajewski and Ziel [14] also analyzed the *EPEX SPOT* ID3 index through a multivariate elastic net regression model. Their difference to Uniejewski et al. [18] is that they included both the one hour and quarterly-hour in their analysis and focused on the variable selection. However, all previous papers disregard the continuous structure of the intraday spot price market as discussed in Sect. 1.1 and 1.2. While the forecast of the MIBEL intraday market covered at least multiple trading intervals [1,13], the German intraday forecast only consider one observation per product [9,14,18]. While this approach might be interesting for a long term perspective (e.g. on the day-ahead market), it does not necessarily reflect the needs on the intraday market. Therefore, this paper introduces a new perspective on the intraday market with the aim to establish a more short-term forecasting horizon.

3   Data Used for the Continuous Intraday Price Forecast

In order to compare different forecasting approaches, we chose the German hourly electricity intraday spot price as endogenous variable. As time horizon, we analyzed the first half of 2018 and thus constructed a data set of the German intraday price from 02.01.2018 till the 30.06.2018. For the model comparison and their predictive power, we decided to use the June of 2018 as test interval and the previous five months as training interval.[2] For each day a total of 24 hourly products were traded which results in a total of 3576 products in the training sample and 720 products within the forecasting sample.[3] To analyze the predictive power in short-term perspective, we chose as dependent variable the volume weighted 15 min averaged trading results of the one hour German continuous intraday market. Through the averaging process, further described in Sect. 4.1, a total of 40 observations per product were created, thus giving a more detailed process of the price development. In order, to predict the dependent variable, multiple features were included in the forecasting process. These variables are the following:

1. Past transaction results which consist of the previous price as well as trading volume and number of trades.
2. Prediction error of the wind forecasts between the day-ahead forecast and the current intraday wind forecast.
3. *EPEX SPOT* M7 orderbook data as well as grid frequency data.
4. Categorical variables describe the weekday, hour of the product and the time till completion for each step.

Due to the fact that some variables have not been frequently used in literature and specific transformations were partly necessary, we shortly discuss the variables in detail.

---

[2] Note that it is planned for future work to consider further training and testing periods to investigate the quality of the models.
[3] Note that two days were excluded from the observation (01.01.2018 & 25.03.2018). The first was excluded due to missing training data of 2017, while the second was generally missing data at the respective day.

3.1 Transaction Data

The transaction data (for 2018) of the continuous intraday market were bought from the EPEX SPOT[4]. The data set contains price and volume information about all executed trades. In addition it also contains information about the execution time (at a one minute resolution) and information about the corresponding product. For the price variable, several aggregation steps were necessary. First, the price was centered through a subtraction of the product's day-ahead price and, due to extreme outliers in 2018, scaled by the standard deviation and a constant c.[5] Second, the price was averaged through a volume weighted mean with two different resolutions. For the endogenous variable an interval of 15 min was used for the averaging. In terms of the exogenous variable, the previous price information were aggregated in two parts. On the one hand, the previous 15 min mean prices were used as look-back variable with a time horizon of 4 h. However, additional research showed that a more detailed description of the 15 min prior to the prediction offered further insights for the forecast. Thus, we included a one minute volume weighted mean for the last 15 min in order to capture the short-term trend. Next to the price variable, we further included the trade volume and count (i.e. number of trades) variable. Due to the fact that we wanted to model a mid-range development with the variables, we chose to aggregate them similar to the first spot price variable. Hence, the volume and count variable were aggregated 15 min mean values recorded for the last four hours.

3.2 EPEX SPOT M7 Orderbook Data

The *EPEX SPOT* M7 orderbook data contains all historic and anonymous orders submitted to the continuous intraday market of the *EPEX SPOT*. This allows a more precise description of the current market situation, as the data also consist of additional information such as available trading volume and the current bid-ask spread for each point in time. In general the information contained in the orderbook data can be divided in ex-ante and ex-post information, see Martin et al. in [12]. Ex-ante information like price and volume are available at the current time of order creation. Ex-post information like the execution price depends on market developments and is not yet available at the time of order creation.[6] In our experiments we combined the information to calculate the current buy and sell prices for 10MWh and 50 MWh which we also included in our models. In Fig. 2 we plotted as an example the current price development as well as the sell and buy price for 10 MWh. The difference between the buy and sell price is also known as the bid-ask spread.

[4] https://www.eex-group.com/eexg/companies/epex-spot.
[5] While many researchers further transform the spot price, e.g. Uniejewski and Weron [19], we were not able to detect improvements in our estimation. Instead, we opted for the simple scaling through a constant c so that 99.7% of the data was within the interval $[-1, 1]$.
[6] The detailed variables in the orderbook data are displayed in the Appendix A based on Martin et al. [12].

Fig. 2. Price development of the 2018-01-18 16:00:00 product with the corresponding buy and sell prices for 10 MWh. The x-axis represents the time t of the day, the y-axis the corresponding price at time t.

3.3 Day Ahead and Short-Term Forecasting of Wind Energy

In order to balance their own balancing group, managers often have to compensate for errors in the day ahead forecast (wind and photovoltaics) on the intraday market. Therefore, in our experiments we use the difference between the actual and the day-ahead wind-power production forecast of Germany as feature.[7] For this purpose, we include the day-ahead and short-term forecast generated by the Fraunhofer IEE into our forecasting framework. The respective forecasts have a resolution of 15 min and are generated every 15 min, and forecast the wind-power production in Germany. For more details about the generation of the day ahead and short-term wind-power production forecasts we refer to Wessel et al. [20]. In our experiments we included all forecast errors (i.e. the difference between the actual and the day-ahead forecast) of the past four hours in our models.

3.4 Grid Frequency

In general, the European interconnected grid requires a grid frequency 50 Hz. Usually, only small deviations from the grid frequency occur, so that only minimal countermeasures by the grid operators are necessary to balance the frequency. In order to intervene, balancing energy is often used to compensate for potential imbalances. Since the use of balancing power is usually associated with high costs, it is possible that changes in the frequency are factored in the current electricity price. Therefore, in our forecast approach we also include the current network frequency, where data is provided by the French Transmission System Operator (TSO) RTE.[8]

---

[7] At the time of writing, we had no adequate photovoltaic feed-in forecast available.

[8] https://clients.rte-france.com/lang/an/visiteurs/vie/vie_frequence.jsp.

4   Price Forecasting Methodology

4.1   A New Framework for Short-Term Energy Price Prediction

In order to forecast the intraday electricity price in a short-term perspective it is necessary to transform the continuous variables to a semi-continuous forecast framework. Consequently, we developed the following forecast framework for all one hour products. In our setting we use historical data to predict the next 15 min volume-weighted average prices as illustrated in Fig. 3.[9] In our regular forecast process, for each product the "forecast-time" was limited to an interval between 4 h until 45 min before delivery time (i.e. the termination) of the product. By reason of their low trading volume (as seen in Fig. 1), observations prior to the 4 h limit were discarded and only used as features. Further, due to the fact that in the last 30 min trading is only allowed within the control areas of the distributors, we excluded this time period as well. Thus, we performed the last forecast 45 min before deliver time in order to ensure that there are no overlaps to the last half hour. As a second step, we aggregate the continuous intraday electricity price with a rolling window. The window generates in 5 min steps volume weighted averages of the next four 15 min blocks. The decision to aggregate the spot price through the volume weighted mean is twofold. First, we ensure that enough observations are summarized, given that especially the early observations are relatively sparse. Moreover, the volume averaging further establishes more stable observations that are less affected by outliers i.e. high prices with a small trading volume. Thus, with 5 min steps in the 3.15h interval we receive a total of 40 observations per product.

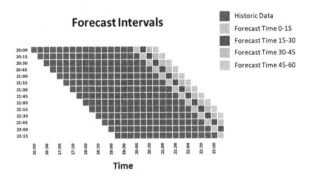

Fig. 3. Forecasting interval of a 00:00 product. The x-axis denotes the time, the y-axis the model run of the forecast (i.e. when the forecast is performed). The first forecast is done four hours before termination (at 20:00), while the last forecast is done 45 min before termination (at 23:15). The dark blue squares symbolize the time horizon of the data points that are used to perform the price forecast for the time symbolized by the remaining squares.

---

[9] In addition, in Sect. 5.2 we extend the prediction interval to a length of four 15 min steps which we further display in Fig. 3 as well.

The third step of the research framework is the inclusion of the features. As already stated in the Sect. 3, the features were included in the analysis through three different groups. As a first group, we consider the price development of the product in the last 15 min with a one minute resolution, to capture the short-term information. This inclusion is based on the idea that last minutes changes in the trading behaviour would translate to the next prediction interval. However, we further wanted to describe a mid-range trend of the spot price. Thus, the second group of variables was included in the framework which consisted of a 4h look-back in 15 min steps. Next to the volume weighted price, this group of variables consisted of the count, volume and orderbook data as well as wind error variable. Lastly, as third group of variables, we included three categorical variables that denoted the hour and weekday of the product as well as the remaining time until the product would end. These variables were included to capture structural dependencies that were valid across all products. After the realization of the described framework, the products were later divided into a training and testing data set, as described in the Sect. 3. Under consideration that the rolling window results in 40 observations per product, we have a total of 28800 observations in the test sample and 143040 observations in the training data set.

4.2 Machine Learning Models

In the following, we briefly summarize the characteristics of the machine learning methods that we applied to the continuous intraday price forecast.

Random Forests (RF). The Random Forests algorithm, introduced by Breiman [2], is an ensemble method that uses a set of (weak) decision trees to build a strong regression model. In the tree building process each tree is trained on bootstrapped samples of the training data. The split of each node is selected on a random subset of d input features.[10] The final prediction is the average of all trees individual predictions.

XG-Boost (XGB). The *Extreme Gradient Boosting* algorithm was first intro-duced by Chen and Guestrin [3] and is a parallel tree boosting that is designed to be "efficient, flexible and portable".[11] In general, the algorithm is based on the *Gradient Boosting* algorithm which is similar to *RF* an ensemble method that combines multiple weak learners to a stronger model. In comparison, the *XG-Boost*-algorithm further improves the framework on the one hand from an algo-rithmic perspective, by enhancing the regularization, weighted quantile sketch-ing and sparsity-aware splitting. On the other hand, the system design was improved through parallelization, distributed tree learning and out-of-core com-putation. Thus, through the combined improvements of both algorithmic and system design, the *XGB* is the next evolutionary step of the *Gradient Boosting*.

---

[10] For regression tasks typically $m/6$ features are selected at random.

[11] https://xgboost.readthedocs.io/en/latest/.

Long Short-Term Memory (LSTM). The *Long Short-Term Memory* was introduced by Hochreiter and Schmidhuber [7] in 1997. A *LSTM* is a special type of *Recurrent Neural Network (RNN)* [17] that can capture long-term dependencies. In contrast to a *RNN*, in practice, a *LSTM* is able to deal with the vanishing and exploding gradient problem. A *LSTM* consists of the four components: memory cell, input gate, forget gate and output gate that interact with each other. Each component is represented by a neural network, where the input gate controls the degree to which level a new information is stored in the memory cell. The forget gate controls the degree to which level an information is kept in the memory cell, and the output gate controls the degree to which level an information is used within the activation function.

5 Evaluation

5.1 Experimental Setting

Given the research data and the respective framework, the three machine learning models *RF*, *XGB* and *LSTM* were applied and optimized for their hyper parameters. Due to the fact that all three models inherit some randomness in their estimation, we aggregate multiple predictions, in order to receive more stable results. The aggregation is implemented by re-estimating the optimized model ten times, forecast based on the test data set and thereafter take the median value of the ten predictions. While the general experimental setting is enforced on all three models, there are still some differences in the structure and the hyper parameter optimization. Accordingly, we shortly summarize the implementation of all three models and further present the baseline as well.

RF and XGB: In terms of the *Random Forests* and the *XG-Boost*, the data was implemented in a regression setting. Because both approaches are in general not able to distinguish time structured data, no further arrangement was necessary. For the hyper parameter optimization of both models, we decided to implement a randomized *Cross-Validation* to ensure that all training data was included in the selection process. Thereafter, the models were repeatedly re-estimated and the median forecast was computed.

LSTM: In the architecture of the *LSTM* model (see Fig. 6 in the Appendix), the features were integrated in three different input gates. For both the 4h mid-range data as well as the 15 min short-term data, the respective features were integrated through separate *LSTM* layers followed by an individual dense layer. In terms of the categorical variable, we decided against the frequently used one-hot encoding and instead chose to embed the variables, based on the results of Guo and Berkhahn [5]. Thus, each variable was first inserted into an embedding layer and then combined with the other two through one dense layer. As a next step, the output of the embedded layers was combined with the two *LSTM* dense layer outputs. Thereafter, the concatenated data is fed through the final two dense layers that summarize the data. For all dense layers, we chose the *Leaky ReLU* activation function with the exception of the last dense layer, were only a linear activation function was implemented. For the hyper parameter optimization, we decided to use the *Hyperband* approach [11], due to its ability to select the best hyper parameters while managing the resources effectively. Given the results of the *Hyperband* optimization, the forecast median was conducted as mentioned before in the experimental setting.

Baseline Models: To examine the forecast quality of our approaches two different and (in practice) commonly used baselines are proposed to guarantee an absolute necessity and minimum performance threshold. The first baseline is the previous observation i.e. the volume weighted mean of the last 15 min which we denote as BL_{15}. As second baseline, we include the last one minute price of the product before the prediction interval, denoted as BL_1.[12] The reason for the decision of two performance thresholds is based on the idea that the models have to compete against both current impulses (BL_1) as well as a more robust price development (BL_{15}) of the product.

5.2   Results and Discussion

Single Step Prediction: Given the aggregated median forecast results, we are now able to analyze and discuss the prediction quality of the three models. For this purpose we primarily evaluate the forecast values based on the *RMSE* error metric. As one can see in Table 1, all three models outperform the BL_{15} baseline by more then 14% as well as the BL_1 baseline by more than 10%. In terms of the overall performance, the *LSTM* shows the best results. However, the results between the *XG-Boost* and the *LSTM* models are relative close which is surprising considering that the *XGB* does not incorporate any time series relationship. Furthermore, when consulting the standard deviation in Table 1 one can see that both shallow learning models are more stable in their prediction. Next, we analyze the performance of the forecast models within the specific time intervals i.e. we evaluate the forecast quality with respect (to the remaining) time till the product's delivery time. The result are visualized in Fig. 4. In this

Table 1. RMSE Results of the respective models and their percentage change in comparison to BL_{15} and BL_1. In the last column we denote the standard deviation of the *RMSE* based on the 10 forecast runs prior to aggregation.

Model	RMSE	$\Delta\%\ BL_{15}$	$\Delta\%\ BL_1$	Standard deviation
BL_{15}	2.3214	0	−3.78	
BL_1	2.2370	3.64	0	
Random Forests	1.9957	14.03	10.78	0.00145
XG-Boost	1.9461	16.17	13.00	0.00766
LSTM	1.9422	16.34	13.18	0.01517

---

[12] As example, for the prediction interval 20:00-20:15 the price of 19.59 is taken as BL_1 baseline and the volume weighted mean of 19:45-20.00 as BL_{15}.

context, some interesting results are visible. First of all, we are able to see that a large proportion of the $RMSE$ errors are induced through the forecast results at the end of the product. While the baseline models BL_{15} and BL_1 jump up to 5.099 and 4.980 in the last forecast step, the machine learning models achieve significant lower values with 3.942 (RF), 3.999 (XGB) and 4.394 ($LSTM$). The most probable explanation for the high rise of the $RMSE$ might be the increase in trading volume as seen in Fig. 1. Furthermore, one can see that all models are frequently able to beat the baseline, with especially good performances at minute 210, 95 and in case of the $LSTM$ also minute 190. It can therefore be assumed that the models will in many cases increase the performance of trading at the intraday market. Lastly, the direct comparison between the $LSTM$ and the XGB in Fig. 4 reveals that the $LSTM$ is generally performing better in the interval between minute 250 and 100, while XGB is showing better results in the last observations. The implication arising from the different performances hints that in the beginning a time series relation might drive the intraday spot price. The advantage of the $LSTM$ is later lost, especially in the last prediction step, thus one might increase the overall performance by combining the two models.

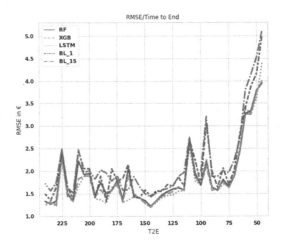

Fig. 4. Display of the $RMSE$ values for different points of time till the end of the product. The x-axis is denoted in minutes till delivery, while the y-axis shows the $RMSE$ in Euro.

The Multi-Step Forecast Extension: With the previous forecast success, we further want to present one possible extension of the forecasting framework. In the prior analysis, we were only interested in a one-step ahead prediction which is most useful for short-term trading. However, for (automated) trading it might be advantageous to increase the forecast horizon in order to model the price trend of the product. Thus, we extended the forecast horizon by three additional 15 min steps. Note that our total forecast interval still ends 30 min before the delivery. This results in fewer forecast intervals in the last steps.

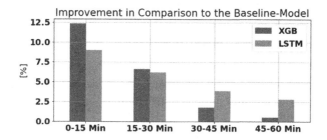

Fig. 5. Bar plot of *RMSE* values for different forecast horizons.

Implementation: For the analysis we applied the two best performing models (*XGB* and *LSTM*).[13] Therefore, in terms of the *XG-Boost* we used a Multi-Output-Regressor structure which basically calculates for each of the four prediction steps an individual model. For the *LSTM* we were able to implement a different approach, because the model is able to predict multiple steps ahead. Hence, a repetition vector prior to the dense layers was embedded and the last two dense layers were enhanced by a time distributed layer.[14,15] Under consideration that the multi-step prediction is only seen as extension, we abstained from a re-optimization of the hyper parameters. Instead we only re-estimate the two models with our median aggregation approach.

Results: With the above-mentioned model-adaptation, the following results are achieved, as displayed in Fig. 5 and Table 2. Overall, both models are still able to outperform the BL_1-baseline in the multi-step ahead forecast. Nevertheless one can observe a weakening of the predictive power the larger the forecasting horizon is. Furthermore, in comparison to Table 1 it is interesting to see that both models perform worse in the first step prediction. For the *LSTM* model, this decline might be induced by the fact that multiple time steps must be optimized. On the other hand, in regard to the second till fourth interval, the *LSTM* is showing a better prediction performance in comparison to the *XGB*. Thus, it might be interesting for future work to identify additional mid-term influencing factors to increase the predictability of the *LSTM*.

---

[13] Since the *RF* had similar values to the *XGB* we skip the analysis of the *RF*.

[14] https://keras.io/api/layers/recurrent_layers/time_distributed/.

[15] The three input gates are kept in the original structure.

Table 2. Multi-step RMSE Results of the *XGB* and *LSTM* model as well as the BL_1.

RMSE	00:00–15:00 min	15:00–30:00 min	30:00–45:00 min	45:00–60:00 min
BL_1	2.2370	3.3569	4.0416	4.7050
LSTM	2.0363	3.1496	3.8851	4.5730
XG-Boost	1.9609	3.1356	3.9687	4.6785

5.3 Future Work

In order to further increase the performance of our algorithms, additional influencing factors of the intraday market must be analyzed in depth and included into the machine learning models. The order *EPEX SPOT* book data, for example, offer high potential here. In this context an interesting research question is which additional features can be gained from the orderbook data to further increase the predictive power. Furthermore, factors such as the forecast error of the photovoltaic feed-in and weather forecasts could be integrated into the models. In addition to more input data, an extension of the forecasting periods should be considered as well. While this paper only examined one month of 2018 it could be of interest, to what extend forecast behaviour is changing with different months and seasons. In the same context, analysis should be conducted on the length of the training data, since it can be assumed that the trading behaviour of (automated) traders changes regularly and "older" data does not adequately reflect the current state of the market. Here, additional fundamental analyses are necessary to obtain meaningful statements. Finally, another aspect is the further development of the proposed *LSTM* model. By integrating *CNN-LSTMs*, new (cnn) features may be generated to obtain better results.

6 Conclusion

In this article we presented the first approach to forecast the electricity price development of individual products at the continuous intraday market of the *EPEX SPOT*. We evaluated the predictive power of two shallow learning algorithm (*XG-Boost* and *Random Forests*) as well as one *LSTM*-based deep learning architecture, complemented by comparing the results with two state-of-the-art baseline models. We show that all considered machine learning models perform significantly better than the baseline models. Our new developed *LSTM*-based model showed the best performance, closely followed by the *XG-Boost*, which had similar results. The remarkable performance of the *XG-Boost* was unexpected, since the *XG-Boost* is not explicitly designed to detect relationships in time series data (in contrast to a *LSTM*). Furthermore we also performed and analyzed a multi-step time series forecast, where we forecast not only the next time step, but the next four. In this context we showed that the quality of the forecast significantly decreases with the longer forecast horizon, however the performance of the baselines was beaten nonetheless.

Acknowledgement. This work was supported as Fraunhofer Cluster of Excellence Integrated Energy Systems *CINES*.

Appendix A   Orderbook Data

(See Table 3)

Table 3. Historic ex-ante and ex-post information available in the M7 orderbook data of the of the continuous intraday market, c.f. [12]. The term "delivery date" is used in this paper equivalently to the term product, and means the time of the delivery start of the corresponding product. The "start validity date" is the time the submitted order is valid from, and the 'end validity date' the time the order is no longer valid. The flag "active order" symbolizes whether an order is active or deactivated. The variable 'side' indicates, if it is a buy or a sell-order. The variables price and volume specify the offered price and volume, in contrast to the 'execution price' and 'execution volume' that define final the price and volume of the respective trade. In this paper we refer to the 'execution price' also with intraday spot price.

Ex-ante	Ex-post
Delivery date	Is executed (yes/no)
Start validity date	End validity date
Active order (Yes/No)	Canceling date
Side (buy/sell)	Execution price
Price	Execution volume
Volume	
ID	

Appendix B   LSTM Architecture

Fig. 6. Structure of the *LSTM* model. On the left side, the three embedding layers process the categorical input. In the middle branch, the look-back variables of the last 4h are feed into the *LSTM* layer. On the right side, the last 15 min are also implemented into the network.

Appendix C   Model Hyper Parameters

In this section we display the hyper parameters for the *RF* in Table 4, the *XGB* in Table 5 and the *LSTM* in Table 6.

Table 4. *RF* hyper parameter

Hyper parameter *XGB*	
N Estimators	1500
Min Samples Split	5
Min Samples Leaf	2
Max Depth	20

Table 5. *XGB* hyper parameter

Hyper parameter *XGB*	
Colsample by Tree	0.8176985004103839
Gamma	0.0979914312095726
Learning Rate	0.04356818667316142
Max Depth	9
N Estimators	184
Subsample	0.7554709158757928

Table 6. *LSTM* hyper parameter

Hyper parameter *LSTM*	
Embedding Layer Output	$\frac{1}{2}$ of Input Shape
LSTM 4h Input	32
LSTM 15min Input	120
LSTM Dropout	0.0001
LSTM Recurrent Dropout	0
LSTM l1,l2 Regularizer	0, 0.000001
LSTM Statefull	False
Dense Layer Embedded	32 units
Dense Layer-4h	40 units
Dense Layer-15min	24 units
Dense Layer Concatenate	50 units
Dense Layer Final	1 units
Leaky ReLU α	0.1
Learning Algorithm	Adam
Learning Rate	0.0005
Batchsize	40

References

1. Andrade, J.R., Filipe, J., Reis, M., Bessa, R.J.: Probabilistic Price Forecasting for Day-Ahead and Intraday Markets: Beyond the Statistical Model (2017)
2. Breiman, L.: Random forests. Mach. Learn. **45**, 5–32 (2001)
3. Chen, T., Guestrin, C.: XGBoost: a scalable tree boosting system. In: Proceedings of the 22nd International Conference on Knowledge Discovery and Data Mining, pp. 785–794 (2016)
4. Frade, P., Vieira-Costa, J.V., Osório, G.J., Santana, J.J., Catalão, J.P.: Influence of wind power on intraday electricity spot market: a comparative study based on real data. Energies **11**, 2974 (2018)
5. Guo, C., Berkhahn, F.: Entity Embeddings of Categorical Variables. arXiv (2016)
6. Gürtler, M., Paulsen, T.: Forecasting performance of time series models on electricity spot markets: a quasi-meta-analysis. Int. J. Energy Sect. Manage. **12**, 103–129 (2018)
7. Hochreiter, S., Schmidhuber, J.: Long short-term memory. Neural Comput. **9**, 1735–1780 (1997)
8. Kath, C.: Modeling intraday markets under the new advances of the cross-border intraday project (XBID): evidence from the german intraday market. Energies **12**, 4339 (2019)
9. Kath, C., Ziel, F.: The value of forecasts: quantifying the economic gains of accurate quarter-hourly electricity price forecasts. Energy Econ. **76**, 411–423 (2018)

10. Kiesel, R., Paraschiv, F.: Econometric analysis of 15-minute intraday electricity prices. Energy Econ. **64**, 77–90 (2017)
11. Li, L., Jamieson, K., DeSalvo, G., Rostamizadeh, A., Talwalkar, A.: Hyperband: a novel bandit-based approach to hyperparameter optimization. J. Mach. Learn. Res. **18**, 6765–6816 (2017)
12. Martin, H.: A Limit Order Book Model for the German Intraday Electricity Market
13. Monteiro, C., Ramirez-Rosado, I.J., Fernandez-Jimenez, L.A., Conde, P.: Short-term price forecasting models based on artificial neural networks for intraday sessions in the iberian electricity market. Energies **9**, 721 (2016)
14. Narajewski, M., Ziel, F.: Econometric Modelling and Forecasting of Intraday Electricity Prices. Journal of Commodity Markets p. 100107 (2019)
15. Oksuz, I., Ugurlu, U.: Neural network based model comparison for intraday electricity price forecasting. Energies **12**, 4557 (2019)
16. Pape, C., Hagemann, S., Weber, C.: Are fundamentals enough? explaining price variations in the german day-ahead and intraday power market. Energy Econ. **54**, 376–387 (2016)
17. Rumelhart, D.E., Hinton, G.E., Williams, R.J.: Learning representations by back-propagating errors. Nature **323**, 533–536 (1986)
18. Uniejewski, B., Marcjasz, G., Weron, R.: Understanding intraday electricity markets: variable selection and very short-term price forecasting using LASSO. Int. J. Forecast. **35**, 1533–1547 (2019)
19. Uniejewski, B., Weron, R.: Efficient forecasting of electricity spot prices with expert and lasso models. Energies **11**, 2039 (2018)
20. Wessel, A., Dobschinski, J., Lange, B.: Integration of offsite wind speed measurements in shortest-term wind power prediction systems. In: 8th International Workshop on Large-Scale Integration of Wind Power into Power Systems, pp. 14–15 (2009)
21. Ziel, F.: Modeling the impact of wind and solar power forecasting errors on intraday electricity prices. In: 2017 14th International Conference on the European Energy Market (EEM), pp. 1–5 (2017)

Neither in the Programs Nor in the Data: Mining the Hidden Financial Knowledge with Knowledge Graphs and Reasoning

Luigi Bellomarini[1]([✉]), Davide Magnanimi[1,2], Markus Nissl[3], and Emanuel Sallinger[3,4]

[1] Banca d'Italia, Rome, Italy
luigi.bellomarini@bancaditalia.it
[2] Politecnico di Milano, Milan, Italy
[3] TU Wien, Vienna, Austria
[4] University of Oxford, Oxford, UK

Abstract. Vadalog is a logic-based reasoning language for modern AI solutions, in particular for Knowledge Graph (KG) systems. It is showing very effective applicability in the financial realm, with success stories in a vast range of scenarios, including: creditworthiness evaluation, analysis of company ownership and control, prevention of potential takeovers of strategic companies, prediction of hidden links between economic entities, detection of family businesses, smart anonymization of financial data, fraud detection and anti-money laundering. In this work, we first focus on the language itself, giving a self-contained and accessible introduction to Warded Datalog+/-, the formalism at the core of Vadalog, as well as to the Vadalog system, a state-of-the-art KG system. We show the essentials of logic-based reasoning in KGs and touch on recent advances where logical inference works in conjunction with the inductive methods of machine learning and data mining. Leveraging our experience with KGs in Banca d'Italia, we then focus on some relevant financial applications and explain how KGs enable the development of novel solutions, able to combine the knowledge mined from the data with the domain awareness of the business experts.

1 Introduction

It is well established that artificial intelligence is playing a key role for digital innovation in the financial domain (globally known as "FinTech"). The increased maturity of AI technology is underpinning an explosive growth of applications that distance themselves from the initial standard FinTech areas (e.g., roboadvisors, P2P lending, crowdfunding, online customer acquisition, mobile wallets, mPOS, personal finance management, cryptocurrencies, private financial planning, etc.) and gain a wider reach, being of growing interest for central authorities

The views and opinions expressed in this paper are those of the authors and do not necessarily reflect the official policy or position of Banca d'Italia.

V. Bitetta et al. (Eds.): MIDAS 2020, LNAI 12591, pp. 119–134, 2021.
https://doi.org/10.1007/978-3-030-66981-2_10

such as national statistical offices, financial intelligence units and central banks. Empowering the business of such central authorities with AI—by either automating mundane tasks or supporting complex decision making—calls for stronger desiderata, in fact with the need for accurate, timely, explainable, accountable and ethical approaches.

A distinguishing aspect of digital innovation at central financial authorities is the opportunity (and indeed the necessity) to exploit a huge amount of domain knowledge, which is a keystone for providing solutions that satisfy the mentioned desiderata. Actually, worldwide, financial authorities are large technical organizations, where critical decisions are based on and combine the experience of teams of expert analysts, whose awareness and sensitiveness cannot be easily mimicked by either pure machine learning solutions or traditional programmatic approaches. In fact, the former may tend to neglect available domain knowledge, with the temptation of indiscriminately relearning it from data, being ineffective, e.g., for compliance checking; the latter, clearly lack the power of inductive approaches, and suffer from costly maintenance, code proliferation, scarce transparency. In other terms, complex financial processes involve humans in the loop for tasks whose knowledge cannot be either reverse-engineered from the data, nor can it be (or has been) effectively encoded in the applications.

So, how to build effective AI-based solutions for complex financial scenarios, able to satisfy the above desiderata, provided that often the domain knowledge cannot be effectively mined from either the enterprise data or the programs?

In this paper we contribute to answering this question by reporting on our experience in mining such financial knowledge from domain experts, programs and data into logic-based Knowledge Graphs (KG) and show we use such technology to build effective AI-enhanced financial applications for Banca d'Italia, the central bank of Italy. We both present elements of our production solutions, which address the mentioned desiderata, and give methodological hints.

Knowledge Graphs are a semi-structured data model, particularly suited for the representation of domains characterized by the presence of many complex entities involved in a dense network of mutual relationships. It turns out this is the case of the financial domain, where concepts of interest (companies, stocks, securities, transactions, people, etc.) are highly intertwined. KGs consist of a *ground extensional component* (the ground truth, e.g., originating from the enterprise data stores), an *intensional component*, that is, the encoding of the domain experience according to a given KRR (*Knowledge Representation and Reasoning*) formalism, and the *derived extensional component*, i.e., the conclusions that can be inferred from the combination of the first two components in the so-called *reasoning process* [9].

For our applications, we rely on VADALOG [13], a language whose core, *Warded Datalog$^\pm$* [10], is a member of the Datalog$^\pm$ family [16] and provides a very good trade-off between computational complexity, being PTIME in data complexity, and expressive power, in fact capturing full *Datalog* [17,19] and being able to express SPARQL queries under set semantics and the entailment regime for OWL 2 QL [10]. In the rest of the work we show how different financial settings can be addressed with VADALOG KGs, adhering to the following methodology:

- We harness the available ground knowledge by mapping the data from the enterprise data stores into facts of the ground extensional component;
- We elicit the business knowledge from domain experts, programs and data and model it in the form of sets of VADALOG rules, so contributing to the construction of the overall KG intensional component;
- We solve business problems by designing software applications that are AI-enhanced, in the sense that they aid experts' work by actively relying on a *Knowledge Graph Management System* (KGMS), which, upon invocation, combines facts and VADALOG rules via logic-based reasoning, while producing new knowledge in the derived extensional components.

Contribution Overview. The remainder of the paper is organized as follows. In Sect. 2 we briefly introduce KGs and VADALOG. Section 3 presents cases where KGs are used to decide on power relationships between companies, with the goal of ultimately assessing who takes decisions and who benefits from financial transactions in a company network. Then, in Sect. 4, we extend these results and apply them to prevent company takeovers; in particular situations of market turbulence, e.g., the ones induced by the recent pandemic, companies are more and more subject to attacks and may fall victim to takeovers, sometimes hostile (in the sense that they happen against the will of the company board), or anyway particularly sensitive especially in the case of companies of strategic relevance. Section 5 introduces a higher-level methodology for link prediction in financial KGs. Section 6 focuses on novel anti-money laundering applications, where KGs are used to harness the knowledge about potentially fraudulent or collusion patterns. Section 7 briefly touches on smart anonymization, which consists in providing forms of statistical confidentiality guarantee for data that are to be published outside the central authority. Finally, Sect. 8 gives some useful references to related work and Sect. 9 concludes the paper.

2   Knowledge Graphs and VADALOG

As we have mentioned, at the core of a Knowledge Graph lies the specific KRR language chosen to represent the intensional component: a modern KGMS aims at languages providing efficient reasoning solutions, balancing pervasive recursion, essential to graph navigation, powerful existential quantification, core to ontological reasoning, and low complexity, to sustain scalability.

The recent resurgence of *Datalog* in academia and industry [2,7,14–16] is turning out to be a key propeller for KGs and reasoners based on logic KRR languages, with many companies and research projects focusing their attention on logic-based KGMS. Fruitful research has been spawned by the need to balance the mentioned complexity vs. scalability requirements. This has been done under different names, which we shall call here *Datalog$^{\pm}$*, the "+" referring to the additional features (including existential quantification), the "−" to restrictions that have to be made to obtain decidability. Many languages within the Datalog$^{\pm}$ family of languages have been introduced and analyzed [5,6,14–16]. Depending on the syntactic restrictions, they achieve a different balance between expressive power and computational complexity.

In our experience in the design of KG-based solutions, we adopt the VADALOG [10] language for the intensional components of the KGs. The VADALOG System [13] is a KGMS whose core revolves around VADALOG. Its architecture (Fig. 1) offers a set of *runtime adapters*, allowing the *core reasoning component* to interact with the ground extensional components, stored in external systems by means of a fact-based (i.e., relational) interface, to infer the facts of the *derived extensional component*.

Fig. 1. The VADALOG system.

A (VADALOG) *rule* is a first-order sentence of the form $\forall \bar{x} \forall \bar{y}(\varphi(\bar{x}, \bar{y}) \rightarrow \exists \bar{z} \psi(\bar{x}, \bar{z}))$,
where φ (the *body*) and ψ (the *head*) are conjunctions of atoms. For brevity, we omit universal quantifiers and denote conjunction by comma. As usual in this context, the semantics of a set of rules is defined by the well-known CHASE procedure. The core of VADALOG is based on *Warded Datalog$^\pm$* [10], a syntactic restriction to Datalog$^\pm$ that guarantees decidability and tractability in the presence of recursion and existential quantification. In terms of expressive power, Warded Datalog$^\pm$ captures full Datalog and *OWL 2* direct semantics entailment regime for *OWL 2 QL*. The language underpinnings are exploited by the reasoner to allow for efficient execution of reasoning tasks [13]. VADALOG augments Warded Datalog$^\pm$ with supplementary features such as aggregation, algebraic operations, and stratified negation.

VADALOG supports *monotonic aggregations*, whose full details can be found in [13]. However, a simpler form of aggregation, which suffices to our ends, is based on *stratified semantics*, where the basic idea for our case is very simple: an aggregation function is computed only when its input operands are completely known. All the scenarios we introduce in this work admit such simplification.

3   Company Control

Let us start by introducing the domain of ownership structures, particularly fit to be modeled as a KG and to be used to solve interesting business questions thanks to reasoning. Ownership structure can be modeled in terms of a *company ownership graph*, i.e., a directed graph where nodes are shareholders (i.e., people and companies, respectively the green and grey nodes in Fig. 2) and edges represent ownership relationships labelled with the fraction of shares that a company or a person x owns of a company y. This constitutes the extensional component of our KG. Two relevant problems follow, which can be both represented with VADALOG rules, so as the KG intensional component.

Analysing the decision power of shareholders is of the essence in such networks. Decision power consists in the ability of a company (or a person) x to induce decision on a company y by controlling the majority of its shares and so the majority of the voting power. A common company control model is the

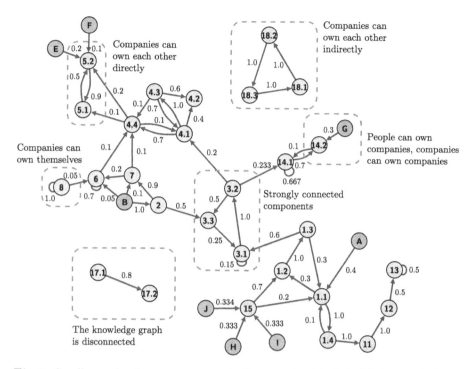

Fig. 2. Small sample of a company graph. Green nodes labeled with letters A-J represent people, while grey nodes labeled with numbers are companies. (Color figure online)

following [17] *A company x controls a company y, if: (i) x directly owns more than 50% of y; or, (ii) x controls a set of companies that jointly (i.e., summing their shares), and possibly together with x, own more than 50% of y.*

In the intensional component of our KG, control edges can be effectively formalized, with the following VADALOG rules:

$$Control(x) \rightarrow Control(x, x) \quad (1)$$

$$Control(x, y), Own(y, z, w), v = msum(w, \langle y \rangle), v > 0.5 \rightarrow Control(x, z) \quad (2)$$

Given that every company has control on itself (Rule (1)), we inductively define control of x on z by iterating over all companies y controlled by x and summing the shares of z that each company y owns (Rule (2)).

Figure 3 shows basic scenarios of company control. In Fig. 3(a) Person A controls Company B by directly owning more than 50% of its share. In Fig. 3(b) Person A does not directly control B; but since she controls 80% of the shares of company C, she controls C. So, the total share of company B that A controls is 61%, which is the sum of the direct 30% ownership of A on B and the 31% ownership of the controlled company C on B. In the end, A controls B. Figure 3(c) shows a more complex form of control: A Person A can control a company B anywhere in the graph even if no direct control exists between them. In fact, in

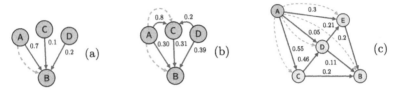

Fig. 3. Sample ownership graphs where *A* controls *B*. Nodes are entities; solid edges are direct ownerships; dashed edges are control relationships. (Color figure online)

this case, Person *A* controls company *B* by controlling (in cascade) some other intermediate companies.

4 Prevention of Company Takeovers

The COVID-19 outbreak poses a host of application settings for the ownership KGs. Relevantly, during market turbulence of crisis times, it is possible that hostile attackers (i.e., people or companies) try to take over some national assets, including those of strategic relevance. A takeover is a change on the public control over a company that moves from a shareholder to another one.

Considering the company control problem presented in Sect. 3, actors can gain the control over target companies either directly, by acquiring the majority of its shares, or indirectly, by gaining the control over a set of intermediate companies such that they jointly own more than 50% of the shares of the target.

Multiple countries have legal tools to protect national strategic companies against foreign takeovers. Italy has the so-called "Golden Power" (GP) that allows the Government to veto individual share acquisition operations that lead to a takeover on strategic national companies. Besides, the Government can intervene to secure companies by acquiring or increasing its participation in the strategic firms (technically, *investment beef-up*) via publicly controlled intermediaries. The application of this tool is by no means trivial. In fact, how can we tell whether a transaction (e.g., a share acquisition operation) is hiding a takeover? What is the minimum amount of share that the Government has to shift under the public control to protect a national strategic company? How can it protect companies when several attackers are acting coordinated and in *collusion*?

These problems lend themselves to being modeled as extensions of the company control setting in the intensional part of the KG (which inherits rules of Sect. 3) and therefore benefit from a *declarative and fully explainable* approach. We developed a rich set of *core reasoning tasks* for 1. *detecting* possible takeover attempts; 2. *suggesting* limits within which GP may be exercised; 3. giving options for *proactively protecting* companies from takeover attempts.

We present three of such problems (Figs. 4, 5 and 6), while a full-detail description can be found in [8]. We consider four types of companies: *trusted* (e.g., public companies or Governmental bodies, pink), *attacker* (e.g., a company out of the national border in the figures, i.e., incorporated or organized under the law of another country), or *target* (e.g., the strategic company to be protected, green); all the others are assumed to be *neutral* (gray). Besides, in all the following figures, solid lines represent already settled ownership relationships while the dashed lines are candidate transactions.

Golden Power Check. It is the basic setting, consisting in the detection of a single transaction that causes a target company to be taken over.

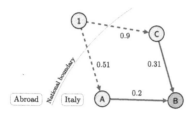

Fig. 4. Example for Golden Power Check. (Color figure online)

Goal. Checking whether an acquisition (of shares, stocks, etc.) causes any target company to become controlled by an attacking company.

Setting. Let T be a set of target companies, V a set of attacking companies, and N a set of neutral companies. Let t be a transaction (e.g., an offer issued by a company x to buy an amount s of shares of a company y), with $x \in V$, $y \in T \cup N$.

Question. Does t cause any company in V to gain control of any company in T?

Insight. If the answer is YES, consider the possibility to block t via GP.

Example. Let us consider the example shown in Fig. 4. Company 1 is the attacker company, while the green node is the target company. Let us also consider the first candidate transaction t_1, where an ownership of 51% if A is going to be acquired by 1. The second candidate transaction is t_2, where an ownership of 90% if C is going to be acquired by 1. Considering only the first candidate transaction, this would give 1 control of A, and hence Company 1 would control the 20% of the shares of Company B. There is no need to block t_1. Now assume that transaction t_1 was processed (i.e., it becomes a solid line), and consider transaction t_2. This would give 1 control of C and hence 31% ownership of B. Together with the ownership of 20% of target company B that 1 already holds, it now has 51% ownership of company B and thus controls it. Transaction t_2 must be blocked using Golden Power. Finally, we remark that had transaction t_2 come before t_1, it would have been fine to process t_2 and block t_1.
This problem can be formalized with the following VADALOG rules:

$$V(x), \neg V(y), Tx(x, y, w) \rightarrow Own(x, y, w) \tag{1}$$
$$V(x), T(y), Control(x, y) \rightarrow GPCheck(x, y) \tag{2}$$

Rule 1 defines that, for the purpose of our analysis, we consider candidate transaction Tx to be virtually applied, i.e., leading to actual ownership even if it has not taken place. Then, Rule 2 computes all companies in V that control at least one company in T. *GPCheck* gives a list of the companies which are possible subject of takeovers caused by the single acquisition of share Tx. If the list is empty, there is no need to employ Golden Power.

Golden Power Protection. As the application of Golden Power comes with political and economic consequences, in many cases is preferable to prevent takeovers by protecting companies in advance avoiding the Golden Power to become a necessity. This third case is called *Golden Power Protection*.

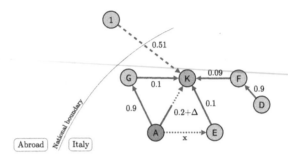

Fig. 5. Example for Golden Power Protection. (Color figure online)

Goal. Computing the share increment needed by trusted companies to prevent takeovers.

Setting. Let T be a set of target companies and V be a set of attacking companies. Let P be a set of trusted companies (such that P is disjoint from V and T).

Question. Which acquisitions of shares of companies in T by companies in P guarantee that no set of transactions t (from x to y, with $x \in V$, $y \in T$) allows any company in V to gain control over one in T?

Insight. Consider the possibility to buy shares of T via P as per the answer to the above question to prevent takeovers (Golden Power is not needed).
The problem can be formalized with the following VADALOG rules:

$$Control(x, y), Own(y, z, w), v = sum(w) \rightarrow PControl(x, z, v) \tag{1}$$

$$P(x), T(y), PControl(x, y, v), v < 0.5 \rightarrow Prot(x, y, 0.5 - v) \tag{2}$$

For every pair of directly or indirectly connected companies, Rule 1 computes the partial control, i.e., the sum of shares of a company that are controlled by the other one. Value v, computed by Rule 2, quantifies the additional direct share ("beef-up") that a trusted company x needs to secure in order to gain control over a target company y.

Cautious Golden Power Check. In the previous examples we have assumed that if only a partial percentage of the ownership of a company is represented, the remaining shares are not relevant. However, we may want to presume the most unfavorable conditions, namely that the non-represented shares are already in the hands of the attackers.

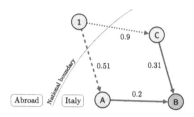

Fig. 6. Example for Cautious Golden Power Check. (Color figure online)

Goal. Checking whether an acquisition (of shares, stocks, etc.) causes any target company to become possibly controlled by an attacking company for which shareholding information is incomplete.

Setting. Let T be a set of target companies, V be a set of attacking companies, and N a set of neutral companies. Let t be a transaction (e.g., an offer issued by a company x to buy an amount s of shares of a company y), with $x \in V$, $y \in T \cup N$.

Question. Assuming that any unassigned share of y is in fact owned by some $v \in V$, does t allow v to gain control of y?

Insight. If the answer is YES, consider the possibility to block t via GP. The following set of VADALOG rules formalize the problem:

$$\neg V(x), \neg V(y), Own(x, y, v), w = sum(v) \rightarrow Assigned(y, w) \qquad (1)$$
$$Assigned(y, w), w < 1 \rightarrow \exists z \; Company(z), V(z), Own(z, y, 1 - w) \qquad (2)$$
$$V(x), \neg V(y), Tx(x, y, w), v = sum(w) \rightarrow Own(x, y, v) \qquad (3)$$
$$V(x), T(y), Control(x, y) \rightarrow GPCCheck(x, y) \qquad (4)$$

We compute the total•amount of currently (un)assigned shares, for each company, with Rule 1. Then, if a neutral or target company has unassigned shares, Rule 2 assigns all of them to an attacker. Finally, Rule 3 computes the integrated ownership and Rule 4 performs the control check.

5   Link Prediction

Company Control, as discussed in Sect. 3 and used in Sect. 4, is only based on financial information, which often does not provide the full picture to understand the actual control relationships holding in practice. For example, family-controlled company groups are hardly detected in case the family membership information is missing, and thus the company is erroneously traced back to multiple centers of interest instead of a single one. Besides family connections, many other hidden relationships exist in different financial settings. A relevant case is that of *close-link* relationships between companies, denoting situations of potential conflict of interest between companies that are financially close and also jointly involved the issuance of asset-backed securities [1].

In this section we introduce VADA-link [3], a link prediction framework where the prediction logic is specified in VADALOG as a KG intensional component that combines logic-based inference on known domain rules and machine learning tasks on the available data.

VADA-Link. We structure the link prediction into two tasks: (i) a double-level *clustering task* to limit the search space (i.e., to avoid an exhaustive comparison of all pairs of people) by using a combination of deterministic rule-based reasoning as well as embedding techniques such as *node2vec*, and (ii) a *multi-class classification task* to assign a link class to a candidate pair of nodes, if existing.

Algorithm 1 presents a high-level overview of our approach. It takes as input a generalized form of a KG, $G = (N, E, \rho, \lambda, \sigma)$, in our setting the company network, and returns another generalized graph with the predicted edges, for example family or close-link relationships, where N is a set of nodes (e.g., companies and people), E is a set of edges (e.g., shareholding edges), $\rho : E \to N^n$ is an incidence function that associates each edge with a set of nodes, in our case $n = 2$, $\lambda : (N \cup E) \to \mathbf{L}$ is a labelling function that associates nodes and edges with a label from a set $\mathbf{L}$ (e.g., $\mathbf{L} = \{C, P, S\}$, where C stands for company nodes, P for person nodes and S for the shareholding edge), and $\sigma : (N \cup E) \times \mathbf{P} \to \mathbf{V}$ associates nodes and edges with properties from $\mathbf{P}$ to a value from a set $\mathbf{V}$ for each property. For example a possible configuration is: each node n has an identifier $x \in P$ with a value $\sigma(n, x)$ and a set of features $f_1, \ldots, f_m$ with value $\sigma(n, f_1), \ldots, \sigma(n, f_m)$, and each edge e has a share amount w with value $\sigma(e, w) \in (0, 1] \subseteq \mathbb{R}$.

The algorithm starts with an embedding-based clustering in Line 5. This function maps the whole graph into a multi-dimensional vector space whose distance, defined by the features and role in the graph topology, measures the similarity of the nodes. Then each resulting cluster K of the embedding is partitioned into more specific clusters B (Line 7). This two-step clustering reduces the search space with a technique similar to *blocking* in record linkage [18]. Then, we consider the nodes in a cluster B to search for links (*Candidate* function) of types in C. In case of a link, a new edge is added with the respective link type (Lines 11–13). The algorithm repeats until now further change occurs.

Clustering. We name the two clustering levels as *GraphEmbedClust* and *GenerateBlocks*. Let us see them in detail. The first level is based on node embedding. In particular, we adapt node2vec in such a way that the distance between nodes reflects the similarity of features as well as neighbourhood relationships.

The second level reduces the search space even further by sub-clustering by feature value, e.g., family names. It is clear that the relationship between features highly depends on domain-specific knowledge. So, we modularize out this aspect and allow for custom implementations of *GenerateBlocks* that provide such functionality in a domain-specific fashion. In most cases, though, such relationships are given by hashing or partitioning techniques.

Algorithm 1. Link prediction algorithm.

Input: $G = (N, E, \rho, \lambda, \sigma)$, C: link classes
Output: $U = (N, E', \rho', \lambda', \sigma)$

```
 1:  U ← G
 2:  changed ← true
 3:  while changed do
 4:      changed ← false
 5:      K ← GraphEmbedClust(U)
 6:      for K ∈ K do                               ▷ First level
 7:          B ← GenerateBlocks(K)
 8:          for B(N_B, E_B, ρ_B, λ_B, σ_B) ∈ B do  ▷ Second level
 9:              for p₁, p₂ ∈ N_B, c ∈ C do
10:                  if Candidate(p₁, p₂, c) and e ∉ E_B then
11:                      add e to E_B
12:                      set ρ_B(e) = (p₁, p₂)
13:                      set λ_B(e, TYPE) = C.
14:                      changed ← true
15:  return U
```

Candidate Prediction. We now consider the classification problem of predicting links. In Algorithm 1 this classification is handled in Line 10 with the *Candidate* predicate. For each link type, specific rules decide when two nodes are considered as connected on the basis of their features; specific link types require the comparison of different features and the combination of the obtained results according to the respective reasoning rules. For instance, two adult people of the same age range, living at the same address, are partners, with some likelihood.

KG-Based Reasoning with VADALOG for Link Prediction. In VADA-link, the intensional component of the ownership KG is extended by the following sets of VADALOG rules: (i) a set of rules encoding a mapping between the ownership graph and a generic abstract graph model where the actual link prediction takes place, in a *schema-independent* but *domain-aware* way, inspired by our experience in mapping and transforming schemas [4]; (ii) a set of rules computing the linkability score, which make use of external machine learning functions in the body; (iii) an overarching set of rules encoding the *GraphEmbedClust* function (Line 5) that returns the top-ranked clusters by combining the different scores. The full sets of rules can be found in [3].

6   Anti-money Laundering

The enterprise Knowledge Graph is a central company asset, where a richer and richer intensional component can be exploited to solve increasingly complex problems. We have seen, e.g., how a central bank can bootstrap a KG from the network of ownership structures and then enhance it with derived control edges, very helpful for instance to prevent takeovers; then, a broader set of relationships, e.g., family connections, open up to more sophisticated tasks like those required for anti-money laundering [11], object of this section.

Money laundering is the process of making illegally-gained money, i.e., dirty money, appear legal and clean by obfuscating its illegal origin. Anti-money laundering aims at preventing and countering such phenomenon. It is a process involving many stakeholders, primarily central banks, financial intelligence units (FIUs) and law enforcement agencies but, importantly, also other actors such as financial intermediaries, brokers, notaries, etc. FIUs collect and analyse suspicious transaction reports (STR) from these actors and produce inquiry reports of the cases as well as hypotheses on the underlying crimes for legal follow-ups. For example, a typical task involves deciding whether a specific financial transaction or case (i.e., a set of transactions with their context, i.e., the network of involved subjects, companies, etc.) meets a definition of *suspicious*, for example provided by some regulatory body, and thus deserves legal follow-up.

We now proceed by example and show an anonymized case from [11], where an STR is evaluated with a *suspiciousness scoring* metric that relies on family links discussed in Sect. 5 and on company control from Sect. 3.

An AML Case Study. Figure 7 shows how we use a KG to combine the available extensional knowledge (solid lines), with the intensional knowledge regarding the scoring method (dashed lines), which requires to reason on company and family relationships. Note that this example is a zoomed-in version of an individual case of a much bigger KG with about 22M nodes and 25.5M links.

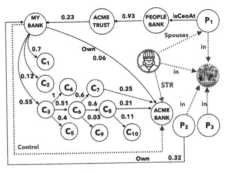

Fig. 7. An Example of AML KG

In the figure we see 10 companies (C1–C10), 4 individuals (P1–P3, and the "guy" we name x, who is part of a family f) and 4 financial institutes (My Bank, ACME Bank, ACME Trust, People Bank). By applying link prediction logic, family f can be singled out and thanks to company control, we learn that f controls Acme Bank. Specifically, f controls My Bank controlling 0.55 of the shares (i.e., 0.32 via P2 and 0.23 via P1, where a CEO edges is assumed as 1) and My Bank controls Acme Bank with 0.52 of the shares via a pyramid shareholding structure of the ten companies to probably obfuscate the connection of the companies.

The STR s from x to Acme Bank reports a loan instance from an individual x having a criminal record. Our goal is to score and explain why s is suspicious.

We rely on a money laundering pattern, well known to the financial intelligence analysts: *A person x who is issuing a loan request to a bank b of which he/she is the ultimate beneficial owner, i.e., a person who ultimately owns or controls the entity on whose behalf a transaction is being conducted (the bank in this case), may intend to launder unclean money via the bank.* In the case of our example, x is not directly controlling the bank, but his family f is controlling it. This means that x may be requesting a fake loan to a controlled bank to justify illegally gained money.

Having already seen the rules to define company control and family relationships, the mentioned pattern can be formalized in the KG intensional component with a single VADALOG rule.

$$STR(x, b, s), Loan(s), In(x, f), Control(f, b) \rightarrow Suspicious(s)$$

In the rule, $In(x, f)$ models that a person x is part of a family f, $Control(z, c)$ states a person or family z controls a company c; $STR(x, b, s)$ is a suspicious transaction report regarding a loan (whence $Loan(s)$), where b is the bank.

A point still remains open: how confident are we of the reasoning conclusion about the suspiciousness of s? It depends on the certainty of the underlying premises, i.e., the results of link prediction and the intrinsic reliability of the anti-money laundering pattern we adopted. All these aspects can be accurately taken into consideration by means of probabilistic extensions of VADALOG [12].

7   Smart Anonymization

As we have seen, Knowledge Graphs are adopted by financial organizations to produce valuable business knowledge for different stakeholders, internal or external to the company, who may have different access and visibility profiles. One meaningful example are the bank supervision data of a central bank, which are needed in full detail by the supervisory bodies, but also serve statistical purposes and have to be publicly disseminated in an anonymized form. In this case, the principle is hiding as much as possible the identity of the involved entities (people, banks, companies, etc.), while preserving the most of statistical properties.

Advanced approaches proposed in the literature, e.g., *differential privacy* [21], tend to be overlooked as too complex for practical application and often financial and statistical organizations develop their own body of best practices and standards. This is the case, for example, of the guidelines [20] of the Italian national statistical office (ISTAT), which has become a de-facto standard for the data anonymity in surveys about families and individuals.

Recent data dissemination initiatives in Banca d'Italia have been following the ISTAT indications and the guidelines have been operationalised in the intensional component of an *anonymization Knowledge Graph*. In particular, a set of Boolean *meta-features* of the involved subjects have been modeled in the extensional component of the KG (Fig. 8), and a set of VADALOG rules in the intensional component have been used to determine confidentiality level of the datasets and allow to take decisions on a threshold basis, namely to prevent identity disclosure and to intervene with ad-hoc anonymization, when needed.

A sample of the mentioned rules defined on a subset of the features (i.e., Italian *fiscal code* (f), *age* (a), *region* (r) and *education* (e)) follows. After standard id-generation (Rule 1) and value normalization (Rule 2), Rule 3 computes the frequency f of an *age-region* pair. Finally, Rule 4 (which in the full KG is combined with many more criteria), increases by one unit the confidentiality level of all people sharing a frequent (therefore "statistically safe") triple.

Features	Dir	N	Rel	Rare	Vis	Tr	Sens
Fiscal code	✗						
Region					✗	✗	
Gender					✗	✗	
Age		✗	✗	✗	✗	✗	
Marital status						✗	
Education						✗	
Hospitalization							✗

Fig. 8. Some meta-features from the ISTAT guidelines. **Dir**: if the feature is "direct", i.e., it directly identifies an individual; **N**: if numeric; **Rel**: when combined ("related") with non-direct features, can disclose confidential information; **Rare**: if specific values can be rare in the dataset; **Vis**: if physically observable in the individual; **Tr**: when traced in other archives, can disclose identities; **Sens**: if sensitive.

$$Person(f, a, r, e), l = 0 \rightarrow \exists i P(i, a, r, e, l) \tag{1}$$
$$P(i, a, r, e, l), 0 < a \leq 20 \rightarrow P(i, 10, r, e, l + 1) \tag{2}$$
$$P(i, a, r, e, l), f = mcount(i)/N \rightarrow FreqR(a, r, f) \tag{3}$$
$$P(i, a, r, e, l), FreqR(a, r, f), f > F \rightarrow \exists x P(i, a, x, e, l + 1) \tag{4}$$

8   Related Work

There is extensive related work for each of the covered topics, as well as some of the topics together. To fit within the scope of this paper, we here refer to the individual works which extensively cover the related work for each topic. The paper [13] covers Knowledge Graphs and VADALOG. Related work for prevention of company takeovers can be found in [8], for link prediction in [3], for anti-money laundering in [11] and smart anonymization in [9].

9   Conclusion

In this paper we presented an overview of the principles and manifold applications of Knowledge Graphs for finance. We highlighted how Knowledge Graphs can be built for representing company control and how such KGs can be utilized to prevent hostile takeovers. We also touched the topics of link prediction, anti-money laundering and smart anonymization, providing both a spectrum of applications to understand the use of KGs in the financial space, as well as illustrating how different AI technologies and reasoning approaches come together.

Many more topics could have been covered. We only touched the surface of how company KGs can be utilized in various finance applications. Beyond company KGs, it is clear that financial Knowledge Graphs can extend to much broader settings and progressively contribute to mine and harness the hidden financial experience into operationalised knowledge.

Acknowledgements. The work on this paper was supported by EPSRC programme grant EP/M025268/1, the EU H2020 grant 809965, and the Vienna Science and Technology (WWTF) grant VRG18-013.

References

1. Guideline (EU) 2018/570 of the ECB (2018). https://www.ecb.europa.eu/ecb/legal/pdf/celex_32018o0003_en_txt.pdf. Accessed 12 Sept 2020
2. Aref, M.: Design and implementation of the LogicBlox system. In: SIGMOD, pp. 1371–1382 (2015)
3. Atzeni, P., Bellomarini, L., Iezzi, M., Sallinger, E., Vlad, A.: Weaving enterprise knowledge graphs: the case of company ownership graphs. In: EDBT (2020)
4. Atzeni, P., Bellomarini, L., Papotti, P., Torlone, R.: Meta-mappings for schema mapping reuse. VLDB **12**(5), 557–569 (2019)
5. Baget, J., Leclère, M., Mugnier, M.: Walking the decidability line for rules with existential variables. In: KR (2010)
6. Baget, J., Mugnier, M., Rudolph, S., Thomazo, M.: Walking the complexity lines for generalized guarded existential rules. In: IJCAI (2011)
7. Barceló, P., Pichler, R. (eds.): Datalog 2.0 2012. LNCS, vol. 7494. Springer, Heidelberg (2012). https://doi.org/10.1007/978-3-642-32925-8
8. Bellomarini, L., et al.: Reasoning on company takeovers during the COVID-19 crisis with knowledge graphs. In: RuleML+RR, vol. 2644, pp. 145–156 (2020)
9. Bellomarini, L., Fakhoury, D., Gottlob, G., Sallinger, E.: Knowledge graphs and enterprise AI: the promise of an enabling technology. In: ICDE, pp. 26–37. IEEE (2019)
10. Bellomarini, L., Gottlob, G., Pieris, A., Sallinger, E.: Swift logic for big data and knowledge graphs. In: IJCAI (2017)
11. Bellomarini, L., Laurenza, E., Sallinger, E.: Rule-based anti-money laundering in financial intelligence units: experience and vision. In: RuleML+RR, vol. 2644, pp. 133–144 (2020)
12. Bellomarini, L., Laurenza, E., Sallinger, E., Sherkhonov, E.: Reasoning under uncertainty in knowledge graphs. In: Gutiérrez-Basulto, V., Kliegr, T., Soylu, A., Giese, M., Roman, D. (eds.) RuleML+RR 2020. LNCS, vol. 12173, pp. 131–139. Springer, Cham (2020). https://doi.org/10.1007/978-3-030-57977-7_9
13. Bellomarini, L., Sallinger, E., Gottlob, G.: The vadalog system: datalog-based reasoning for knowledge graphs. PVLDB **11**(9), 975–987 (2018)
14. Calì, A., Gottlob, G., Kifer, M.: Taming the infinite chase: query answering under expressive relational constraints. J. Artif. Intell. Res. **48**, 115–174 (2013)
15. Calì, A., Gottlob, G., Lukasiewicz, T.: A general Datalog-based framework for tractable query answering over ontologies. J. Web Sem. **14**, 57–83 (2012)
16. Calì, A., Gottlob, G., Lukasiewicz, T., Marnette, B., Pieris, A.: Datalog+/-: a family of logical knowledge representation and query languages for new applications. In: LICS, pp. 228–242 (2010)
17. Ceri, S., Gottlob, G., Tanca, L.: Logic programming and databases. Springer (2012)
18. Christen, P.: Data Matching - Concepts and Techniques for Record Linkage, Entity Resolution, and Duplicate Detection. Springer, Data-centric systems and applications (2012)
19. Huang, S.S., Green, T.J., Loo, B.T.: Datalog and emerging applications: an interactive tutorial. In: SIGMOD (2011)

20. ISTAT, Virgili, L., Foschi, F.: La produzione di MFR e mIcro.STAT (2013). https://www.istat.it/en/files/2013/12/Linee-guida-MFR-e-mIcro.STAT_.pdf. Accessed 12 Sep 2020
21. Kulkarni, T., Cormode, G., Srivastava, D.: Answering range queries under local differential privacy. CoRR, abs/1812.10942 (2018)

Exploring the Predictive Power of News and Neural Machine Learning Models for Economic Forecasting

Luca Barbaglia, Sergio Consoli[⊠], and Sebastiano Manzan

European Commission, Joint Research Centre, Directorate A-Strategy, Work Programme and Resources, Scientific Development Unit, Via E. Fermi 2749, 21027 Ispra, VA, Italy
sergio.consoli@ec.europa.eu

Abstract. Forecasting economic and financial variables is a challenging task for several reasons, such as the low signal-to-noise ratio, regime changes, and the effect of volatility among others. A recent trend is to extract information from news as an additional source to forecast economic activity and financial variables. The goal is to evaluate if news can improve forecasts from standard methods that usually are not well-specified and have poor out-of-sample performance. In a currently ongoing project, our goal is to combine a richer information set that includes news with a state-of-the-art machine learning model. In particular, we leverage on two recent advances in Data Science, specifically on Word Embedding and Deep Learning models, which have recently attracted extensive attention in many scientific fields. We believe that by combining the two methodologies, effective solutions can be built to improve the prediction accuracy for economic and financial time series. In this preliminary contribution, we provide an overview of the methodology under development and some initial empirical findings. The forecasting model is based on DeepAR, an auto-regressive probabilistic Recurrent Neural Network model, that is combined with GloVe Word Embeddings extracted from economic news. The target variable is the spread between the US 10-Year Treasury Constant Maturity and the 3-Month Treasury Constant Maturity (T10Y3M). The DeepAR model is trained on a large number of related GloVe Word Embedding time series, and employed to produce point and density forecasts.

Keywords: Economic and financial forecasting · Neural time series forecasting · Deep learning · Recurrent neural networks · Long short-term memory networks · Word embedding · News analysis

1 Introduction

Monitoring the current and forecasting the future state of the economy is of fundamental importance for governments, international organizations, and central banks. Policy makers require timely macroeconomic information in order

---

Authors listed in alphabetical order.

© The Author(s) 2021
V. Bitetta et al. (Eds.): MIDAS 2020, LNAI 12591, pp. 135–149, 2021.
https://doi.org/10.1007/978-3-030-66981-2_11

to design effective policies that can foster economic growth and preserve societal well-being. However, they rely on economic indicators produced by statistical agencies that are released at low frequencies (e.g., monthly or quarterly), with considerable delays, and that are often subject to substantial revisions. With such an incomplete information set in real-time, the forecasts produced by economists are very uncertain and inaccurate even when forecasting the current economic situation as well as the future, thus making the task extremely challenging. Moreover, in a global interconnected world, shocks and changes originating in one economy could move quickly to other economies affecting productivity levels, job creation and welfare in different geographic areas. However, greater interdependence also means that the current and future conditions of a market are linked to instabilities and extreme events originated abroad. All these factors make the economic forecasting task extremely difficult, both in the short and in the medium-long run.

In this context, economists and researchers can leverage on the rapid advances in information and communications technology experienced in the last two decades, which have produced an explosive growth in the amount of information available leading to the era of *Big Data* [17]. Novel and alternative data sets can potentially contribute greatly to help us in monitoring and forecasting economic activity given its timeliness and its economic relevance. A major source of such information is represented by news text, since it discusses important events, economic and financial news releases, and experts opinions, among others, that can serve as a basis for economic and financial decisions [5]. In addition, news affect consumers' perception of the economy through three channels. First, the news media convey the latest economic data and professionals' opinion to consumers. Second, consumers receive a signal about the economy through the tone and volume of economic reporting. Last, the greater the volume of news about the economy, the greater the likelihood that consumers will update their expectations about the economy [7].

In a currently on-going research project, we aim to explore the predictive power of news for forecasting of economic and financial time series by leveraging on the recent advances in Data Science, specifically on Word Embedding [6,18,20] and Deep Learning [3,15,21,22] models. Word Embedding models represent the contextual information of a given corpora and capture syntactic and semantic information with respect to the data set used for building the embeddings. Following a specific NLP information extraction pipeline, which we describe in Sect. 4, we extract sentences referring to a specific economic aspect from the news media. Then, from those sentences, we derive signals corresponding to the Word Embedding of the extracted sentences. For example, if we deal with daily economic time series data, the Word Embedding signals are calculated on the sentences extracted each day from the news text. In our on-going research activity we aim to compare the predicting capabilities of the most popular Word Embedding methods, from the widely used Word2Vec [18] and GloVe [20] models, to the most recent context-dependent models like BERT [6] or GPT-3 [4]. In this contribution we only consider Glove embeddings. The goal is to extract the hidden information embedded in economic news to provide useful predictive signals that can be used as additional features to improve the accuracy of economic forecasts [9].

In addition, we use a novel forecasting model based on Deep Learning [3, 14, 15] for addressing the economic forecasting task. In particular, in this work we rely on DeepAR [22], a powerful neural forecasting methodology that produces accurate probabilistic forecasts, based on training an auto-regressive Recurrent Neural Network (RNN) model on a large number of related time series, which in our case are the Word Embedding signals. We believe that by combining these two strong Data Science methodologies, that is Word Embedding and Deep Learning, effective solutions can be built to improve the accuracy of prediction tasks for economic and financial time series.

In this contribution we provide an overview of the methodology under development. We report on some preliminary findings on the use-case application of DeepAR along the Word Embedding extracted by a GloVe pre-trained model from United States (US) news ranging from January 1982 to September 2019, with the goal of predicting the future values of the US 10-Year Treasury Constant Maturity Minus 3-Month Treasury Constant Maturity (T10Y3M) time series given its past values.

2   Background

Information encoded in text is a rich complement to the more structured kinds of data traditionally used in empirical research [10]. In recent years, we have seen an intense use of textual data in different areas of research. The idea consists of transforming strings of raw text into numeric variables, and then use them as predictors in different models. News articles, in particular, represent a relevant data source to model economic and financial variables, and several studies have already explored this additional source of information. For a recent overview on the application of text analysis in economics and finance the reader is also referred to [1, 10].

On the other end, there is a vast literature on the use of Deep Learning in the context of time series forecasting, e.g. see [14, 21]. For a survey the reader is referred to [3, 13, 15]. Neural Networks (NNs) in forecasting have been typically applied to individual time series, i.e. a different model is fitted to each time series independently [23]. Although it is fairly straightforward to use classic Multilayer Perceptrons NNs (aka MLPs) on large data sets, its use on medium-sized time series is more difficult due to the high risk of over-fitting. Classical MLPs can be adapted to address the sequential nature of the data by treating time as an explicit part of the input. However, such an approach has some inherent difficulties, namely the inability to process sequences of varying length and to detect time invariant patterns in the data. A more direct approach is to use recurrent connections that connect the neural networks hidden units back to themselves with a time delay. This is the principle at the base of Recurrent Neural Networks (RNNs) [15, 21], which are NNs specifically designed to handle sequential data that arise in applications such as time series, natural language processing and speech recognition. In finance, for example, the authors in [16] developed a multi-task RNN with high-order Markov random fields to predict stock price movement direction based upon a single stock's historical records together with its correlated stocks.

Although RNNs have been widely used in practice, it turns out that training them is quite difficult given that they are typically applied to very long sequences of data. A common issue while training very deep neural networks by gradient-based methods using back-propagation is that of vanishing or exploding gradients which renders learning impossible. Long Short-Term Memory Networks (LSTMs) were proposed [12, 14] to address this problem. Instead of using a simple network at each time step, LSTMs use a more complicated architecture composed of a cell and gates which control the flow of input to the cell as well as decide what information should be kept inside the cell and what should be propagated to the next time step [14]. The cell has a memory state which is propagated across time along with the output of the LSTM unit, which is itself a function of the cell state. Unlike the output of the LSTM unit, the cell state undergoes minimal changes across time, thus the derivative with respect to the cell state does not decay or grow exponentially [11]. Consequently, there is at least one path where the gradient does not vanish or explode making LSTMs suitable for processing long sequences. For details on the working mechanisms behind RNNs and LSTMs we suggest the reader to go through the online tutorial in [19].

Recently, in [22] the authors have proposed DeepAR, an RNN-based forecasting model using LSTM or GRU cells, the latter being a simplification of LSTMs that do not use a separate memory cell and may result in good performance for certain applications. At each time step, DeepAR takes as input the previous time points and covariates, and estimates the distribution of the value of the next period. This is done via the estimation of the parameters of a pre-selected parametric distribution (e.g. negative binomial, student t, gaussian, etc.)[1]. Training and prediction follow the general approach for auto-regressive models [22]. One feature makes this forecasting setting appealing: in probabilistic forecasting one is interested in the full predictive distribution, not just a single best realization, making the analysis more robust and reducing uncertainty in the downstream decision-making flow. In addition to providing more accurate forecasts, DeepAR has also other advantages compared to classical approaches and other global methods [22]: (i) As the model learns seasonal behavior and dependencies on given covariates across time series, manual feature engineering is drastically minimized; (ii) DeepAR makes probabilistic forecasts in the form of Monte Carlo samples that can be used to compute consistent quantile estimates for all sub-ranges in the prediction horizon; (iii) By learning from similar items, DeepAR is able to provide forecasts for items with little history, a case where traditional single-point forecasting methods fail; (iv) DeepAR does not assume Gaussian noise, but can

---

[1] In our case, we considered a student t-distribution in order to account for the fat-tail characteristic of financial returns.

incorporate a wide range of likelihood functions, allowing the user to choose one that is appropriate for the statistical properties of the data.

3   Data

The recent works in economics and finance on the application of text analysis from social media and news generally suffer from a limited scope of historical financial news available, and from the limitation of the analysis to short texts only (e.g. usually tweets or news headlines, see e.g. [1,8]). In our study we consider a long time period and analyse the entire text contained in the news articles. The source of economic news was obtained from a commercial provider[2]. The data set consists of several million articles, full-text, from January 1982 until September 2019 (approximately 40 years) for the following US outlets: The New York Times, The Wall Street Journal, The Washington Post, The Dallas Morning News, The San Francisco Chronicle, and the Chicago Sun-Times. The economic variable of interest is the spread between the yield on 10-year T-bonds and 3-month T-bill, denoted by T10Y3M, which is obtained from the Saint Louis Fed FRED repository[3].

4   Information Extraction from News

In this section we describe the NLP information extraction pipeline used to extract sentences from the news text that refer to specific economic aspects, which we then use to derive the Word Embedding to be used in the forecasting exercise. The method works as follow. Suppose we want to extract from our news data the embeddings referred to a specific economic concept, for instance *industrial production*. First, its economic synonyms are derived from SPARQL queries of the *World Bank Group Ontology*[4]. For example, for *industrial production*, the economic synonyms that are obtained are: manufacturing; industrial output; secondary sector; industry productivity; manufacturing development; industrial growth; manufacturing productivity; etc. Given the goal of forecasting the T10Y3M time series, we used search keywords that are broadly related to the economy and to monetary and fiscal policy (the complete list can be found in the Appendix).

Afterwards, we employ a rule-based procedure that builds on the linguistic features of the *spaCy* Python library[5]. The NLP pipeline relies on the *en_core_web_lg* model of spaCy[6], an English multi-task Convolutional Neural

---

[2] Dow Jones DNA: Data, News and Analytics Platform: https://www.dowjones.com/dna/.

[3] https://fred.stlouisfed.org/series/T10Y3M.

[4] World Bank Group Ontology, available at: http://vocabulary.worldbank.org/thesaurus.html.

[5] spaCy: Industrial-Strength Natural Language Processing. Available at: https://spacy.io/.

[6] https://spacy.io/models/en.

Network trained on OntoNotes and GloVe vectors trained on Common Crawl. The role of *spaCy* is to provide structured information about the text analyzed in the form of word vectors, context-specific token vectors, Part-of-speech (POS) tags, dependency parse and named entities. The following NLP steps are performed:

- *Tokenization & lemmatization:* News text is split into meaningful segments (*tokens*), considering noninflected form of the words (*lemmas*) in the text taken from WordNet[7].
- *Named Entity Recognition:* Named-entity mentions in the news text are located and classified, including locations, organizations, time expressions, quantities, monetary values, etc.
- *Most frequent location:* Heuristic procedure which assigns the location to which a sentence is referring, as its most frequent named-entity location detected in the sentence text.
- *POS tagging:* News text is parsed and tagged using spaCy's statistical model. We loop over the part-of-speech tags, stopping when our search concept, or one of its synonyms, is found.
- *Dependency Parsing:* After our search concept is found in a sentence, we loop over the available dependency parsing tree of the sentence. In particular we navigate over the neighbouring tokens of the discovered search concept by employing a rule-based approach leveraging on the syntactic dependency parsing tree. In this way chunks of terms related to our search concept are constructed.
- *Tense detection:* Heuristic procedure used to detect the tense of the constructed terms chunks extracted from the news and related to our search concept through our rule-based approach.

After the NLP pipeline has produced the chunks of terms related to our search concepts, we merge the GloVe Word Embeddings by averaging each embedding features, thus producing a unique vector of embeddings representing the extracted terms in that sentence. Similarly, all the extracted embeddings of the same frequency period of the time series considered (daily in our case) are merged together producing a unique vector of embeddings for each period. These represent the signals ordered in time that will be used as covariates for the forecasting exercise in our T10Y3M case study.

5 Preliminary Findings

In this section we show our preliminary findings on the application of DeepAR to the forecasting of the T10Y3M time series, augmented by the extracted Word Embedding from the US economic news. Given that the T10Y3M daily time series (see Fig. 1, top) is a highly persistent and non-stationary process, we take

---

[7] WordNet, A Lexical Database for English. Available at https://wordnet.princeton. edu/.

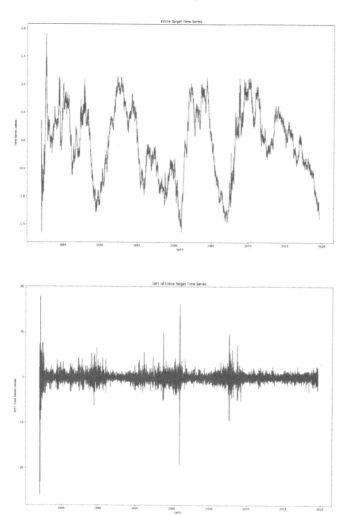

Fig. 1. Time series of the yield spread between the US 10-Year Treasury Constant Maturity and the 3-Month Treasury Constant Maturity (T10Y3M; top panel), and the change of T10Y3M defined as the log-difference between consecutive observations (bottom panel).

its first log-difference and obtain a stationary series of daily changes (see Fig. 1, bottom). Forecasting the yield spread differences is an extremely challenging task, as the series behaves similarly to a random walk process. The time ranges from January 1st 1982 to August 30th 2019. Missing values for the target time series and covariates, mainly related to weekends and holidays, are dropped from the analysis, giving a final number of 9, 418 data points for the whole target time series.

All training, validation, and test data are historical values that have been smoothed using a logarithmic transformation and scaled on training data only. We have opted for a robust scaling of the variables by using statistics robust to the presence of outliers. That is, we remove the median to each time series, and the data are scaled according to the quantile range between the first quartile and the third quartile. Data standardization is a common requirement in the estimation of many machine learning models. Typically this is done by removing the mean and scaling to unit variance. However, outliers can often influence the sample mean/variance negatively. In such cases, as in ours, the median and the inter-quartile range provide better results. Centering and scaling happen independently on each feature by computing the relevant statistics on the training sets. These pre-processing approaches have been used throughout the analysis due to better experimental results relative to other smoothing and scale transformations. We employ a rolling window for training and validation, with a window length equal to half of the full sample, that is 4,709 data points. For each window, we make one step-ahead forecasts.

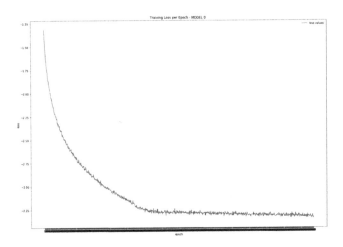

Fig. 2. Example of negative log-likelihood training loss values for the first trained DeepAR model.

The DeepAR model is implemented by adopting Gluon Time Series (GluonTS) [2][8], an open-source library for probabilistic time series modelling that focuses on deep learning-based approaches. The library is available in Python and relies on Gluon, which is the joint AWS/Microsoft open-source deep learning solution interfacing Apache MXNet[9]. The DeepAR parameters are set experimentally to 2 RNN layers, each having 40 LSTM cells, and using a learning rate

---

[8] Available at: https://gluon-ts.mxnet.io/#gluonts-probabilistic-time-series-modeling.

[9] Available at: https://mxnet.apache.org/.

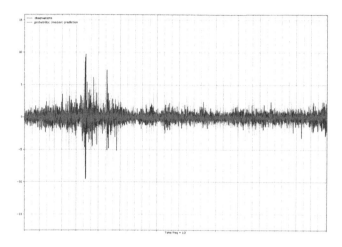

Fig. 3. Median forecasts (green) and observations for the T10Y3M series (blue) for the entire forecasting period. (Color figure online)

equal to 0.001. As mentioned earlier, the employed training loss is the negative log-likelihood, and the probability distribution to draw the probabilistic forecasts is the student t-distribution. We set also a re-training step for the model equal to 7 d, meaning that every 7 consecutive data points the DeepAR model is completely retrained. As concerned the number of epochs for each training, we choose experimentally the value of 500 epochs, which delivers convergence of the training loss (e.g., see in Fig. 2 the training loss values for the first trained model).

The whole experiment is run in parallel on 40 cores at 2.10 GHz each into an Intel(R) Xeon(R) E7 64-bit server having overall 1 TB of shared RAM. The complete experiment required around 20 h of computation time.

Figure 3 shows the observations for the T10Y3M series (blue line) together with the median forecast (dark green line) and the confidence intervals in lighter green (i.e., dark green area = 50% confidence interval; light green area = 90% confidence interval"). To better visualize the differences between observed and predicted time series, we report the same plot on a smaller time range (50 d) in Fig. 4. A qualitative analysis of the figure suggests that the DeepAR forecasts do a reasonable job at capturing the variability and volatility of the time series. Forecasting an interval rather than a point is an important feature of the process since it provides an estimate of the uncertainty involved in the forecast which allows downstream decisions based to account for such uncertainty.

For a quantitative evaluation of the forecasts in the test set, we compute a number of commonly used metrics, such as the mean absolute scaled error (MASE), the symmetric mean absolute percentage error (sMAPE), the root mean square error (RMSE), and the (weighted) quantile losses (wQuantileLoss),

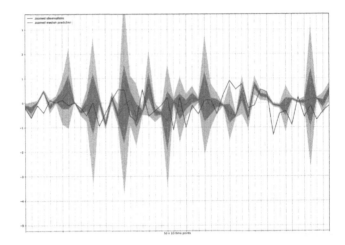

Fig. 4. Probabilistic forecasts (green) and observations for the T10Y3M series (blue) for the first 50 d in the testing period. The green continuous line shows the median of the probabilistic predictions, while lighter green areas represents higher confidence intervals (dark green = 50% confidence interval; light green = 90% confidence interval"). (Color figure online)

that is the quantile negative log-likelihood loss weighted with the density. In particular, DeepAR obtains the following *in-sample results*:

- MASE = 0.42,
- sMAPE = 0.89,
- RMSE = 0.77,
- wQuantileLoss[0.1] = 0.24,
- wQuantileLoss[0.3] = 0.55,
- wQuantileLoss[0.5] = 0.65,
- wQuantileLoss[0.7] = 0.54,
- wQuantileLoss[0.9] = 0.24.

The *out-of-sample results* are instead:

- MASE = 0.75,
- sMAPE = 1.46,
- RMSE = 1.02,
- wQuantileLoss[0.1] = 0.82,
- wQuantileLoss[0.3] = 1.09,
- wQuantileLoss[0.5] = 1.16,
- wQuantileLoss[0.7] = 1.11,
- wQuantileLoss[0.9] = 0.84.

As expected the results worsen passing from the in-sample to the out-of-sample setting. However, the gap looks acceptable showing a good generalization of the trained model. Moreover, the model performs better at high (0.9) and low

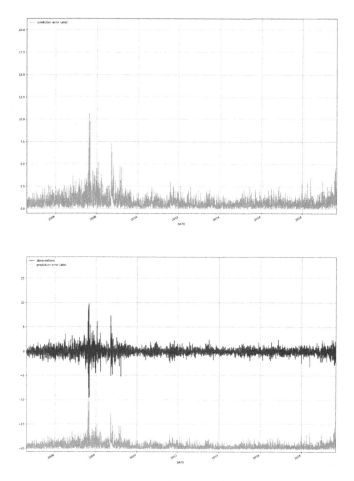

Fig. 5. Median absolute forecast error (MAFE) (top panel), and real T10Y3M observations (blue) against MAFE (orange) (bottom panel). (Color figure online)

(0.1) quantiles, where it obtains lower weighted quantile losses. However, looking at Fig. 5, which shows the median absolute forecast error (MAFE, in orange) against the real T10Y3M observations (in blue), we see that the model is performing poorly during the crisis period (2007–2009), where the series presents clusters of high volatility. In future work, we aim to improve the performance of the algorithm by tweaking our NLP pipeline over the economic news by considering different economic searches and Word Embedding models, and by improving DeepAR's forecasting results by changing architecture and/or hyperparameters.

To evaluate the forecasting performance of DeepAR we can compare the forecasting metrics against those produced by other models. For example, we can produce forecasts using a simple moving average (MA) and a naïve method (NM). With the moving average method, the forecasts of all future values are equal to the average (or "mean") of the historical data. If we let the historical data be denoted by $y_1, ..., y_T$, then we can write the forecasts as $\hat{y}_{T+h|T} = \bar{y} = (y_1 + ... + y_T)/T$. The notation $\hat{y}_{T+h|T}$ is a short-hand for the estimate of y_{T+h} based on the data $y_1, ..., y_T$. In our case we have chosen $T = 7$, that is we do a one-week moving average on the T10Y3M time series. For the naïve forecasts, instead, we simply set all forecasts to be the value of the last observation for our target (i.e. the log-difference of T10Y3M). That is, $\hat{y}_{T+h|T} = \hat{y}_T$. Because naïve forecasts are optimal when data observations follow a random walk, these are also called random walk forecasts. The naïve method works well for many economic and financial time series, as is the case with our T10Y3M data. Table 1 reports the out-of-sample results for the three methods listed above. There is a clear superiority of the DeepAR algorithm with respect to the two other approaches. An extensive computational analysis where the algorithm will be compared with other state-of-the-art forecasting approaches is the goal of future work.

Table 1. Out-of-sample forecast evaluation for the *DeepAR*, *MA*, and *NM* models in terms of MASE, sMAPE, RMSE, and wQuantileLoss$_{mean}$ error metrics.

metrics	DeepAR	MA	NM
MASE	0.75	0.80	0.90
sMAPE	1.46	1.55	1.47
RMSE	1.02	1.04	1.30
wQuantileLoss$_{mean}$	0.88	0.93	1.39

6 Conclusions

In this contribution we provide some early results of a currently on-going project aimed at exploring the predictive power of news for economic and financial time series forecasting. The novelties of the approach is that we use Word Embeddings as features which we use in the DeepAR neural forecasting method. In this initial experiment, we forecast the yield spread between the US 10-Year Treasury Constant Maturity and the 3-Month Treasury Constant Maturity (T10Y3M). The Word Embeddings are calculated based on the GloVe model and extracted from US economic news. After providing an overview of the methodology under current development, we report some preliminary results on this use-case application, showing satisfactory performance of the devised approach for such a challenging task. Information extracted from news looks relevant for the forecasting exercise of this economic variable.

Certainly further extensive research still needs to be done. We notice that DeepAR attains a poor forecasting performance during the crisis period (2007–2009), where the T10Y3M suffered from high volatility. We aim to improve the performance of the algorithm by tweaking our NLP pipeline over the economic news by considering different economic searches. Furthermore, while in this contribution we only use Glove Word Embedding, in the future we will perform a comparison with other models, like the popular Word2Vec and the recent context-dependent BERT model. Moreover, we will try to improve the performance of the implemented DeepAR model by changing architecture and optimizing its hyperparameters. Interpretability of the model by using, e.g., computed Shapley values, will be object of future investigation.

To conclude, our approach that combines Word Embedding and Deep Learning seems a promising direction to follow in order to improve the forecast accuracy of economic and financial time series. We believe that by combining these two methodologies, effective solutions can be built to improve effectiveness for this type of prediction tasks.

Appendix: List of Keywords Used for the Information Extraction Pipeline

The considered single and compounded keywords were the ones belonging to the following two (lemmatized) terms sets:

- single terms:
  "dividend, earning, nasdaq, bank, derivative, lending, borrowing, economy".
- compounded terms:
  "financial market, corporate finance, stock market, capital market, currency market, derivative market, bond market, exchange market, stock price, s&p 500, dow jones, future trading, option trading, financial investor, dow jones industrial average, stock index, equity market, wall street, central bank, fomc, federal reserve, money supply, monetary policy, federal fund, base rate, interest rate, three-month treasury, threemonth treasury, ten-year treasury, ten year treasury, treasury yield, yield curve, yield spread, quantitative easing".

In addition, the considered search keywords contained also the pairwise combinations of the terms in the following two sets:

- single terms combination:
  "banking, financial".
- compounded terms combination:
  "sector, commercial, investment".

The locations and organizations used for restricting the searches to US articles only were:

- locations and organizations:
  "america, united states, columbia, land of liberty, new world, u.s., u.s.a., us of a, usa, us, land of opportunity, the states, fed, federal reserve board, federal reserve, census bureau, bureau of economic analysis, treasury department, department of commerce, bureau of labor statistics, bureau of labour, department of labor, open market committee, bea, bureau of economic analysis, bis, bureau of statistics, board of governors, congressional budget office, cbo, internal revenue service, irs".

References

1. Agrawal, S., Azar, P., Lo, A.W., Singh, T.: Momentum, mean-reversion and social media: evidence from StockTwits and Twitter. J. Portfolio Manag. **44**, 85–95 (2018)
2. Alexandrov, A., et al.: Probabilistic time series models in python. J. Mach. Learn. Res. **21**, 1–6 (2020)
3. Benidis, K., et al.: Neural forecasting: Introduction and literature overview. CoRR, abs/2004.10240, 21, pp. 1–6 (2020)
4. Brown, T.B., et al.: Language models are few-shot learners. CoRR, abs/2005.14165 (2020). https://arxiv.org/abs/2005.14165
5. Chang, C.-Y., Zhang, Y., Teng, Z., Bozanic, Z., Ke, B.: Measuring the information content of financial news. In: Proceedings of COLING 2016– 26th International Conference on Computational Linguistics, pp. 3216–3225 (2016)
6. Devlin, J., Chang, M.-W., Lee, K., Toutanova, K.: BERT: pre-training of deep bidirectional transformers for language understanding. In: Proceedings of NAACL-HLT 2019 - The 17th Annual Conference of the North American Chapter of the Association for Computational Linguistics: Human Language Technologies, vol. 1, pp. 4171–4186 (2019)
7. Doms, M., Morin, N.J.: Consumer sentiment, the economy, and the news media. Finance and Economics Discussion Series 2004-51, Board of Governors of the Federal Reserve System (U.S.) (2004)
8. Dridi, A., Atzeni, M., Recupero, D.R.: FineNews: fine-grained semantic sentiment analysis on financial microblogs and news. Int. J. Mach. Learn. Cybern., 1–9 (2018)
9. Fabbi, C., Righi, A., Testa, P., Valentino, L., Zardetto, D.: Social mood on economy index. In: XIII Conferenza Nazionale di Statistica (2018)
10. Gentzkow, M., Kelly, B., Taddy, M.: Text as data. J. Econ. Lit. (to appear) (2019)
11. Gers, F.A., Eck, D., Schmidhuber, J.: Applying LSTM to time series predictable through time-window approaches. Lect. Notes Comput. Sci. **2130**, 669–676 (2001)
12. Hochreiter, S., Schmidhuber, J.: Long short-term memory. Neural Comput. **9**, 1735–1780 (1997)
13. Januschowski, T., Gasthaus, J., Wang, Y., Salinas, D., Flunkert, V., Bohlke-Schneider, M., Callot, L.: Criteria for classifying forecasting methods. Int. J. Forecast. **36**(1), 167–177 (2020)
14. Lai, G., Chang, W.-C., Yang, Y., Liu, H.: Modeling long- and short-term temporal patterns with deep neural networks. In: 41st International ACM SIGIR Conference on Research and Development in Information Retrieval, SIGIR 2018, pp. 95–104 (2018)
15. Lecun, Y., Bengio, Y., Hinton, G.: Deep learning. Nature **521**(7553), 436–444 (2015)

16. Li, C., Song, D., Tao, D.: Multi-task recurrent neural networks and higher-order Markov random fields for stock price movement prediction. In: Proceedings of the ACM SIGKDD International Conference on Knowledge Discovery and Data Mining, pp. 1141–1151 (2019)
17. Marx, V.: The big challenges of Big Data. Nature **498**, 255–260 (2013)
18. Mikolov, T., Chen, K., Corrado, G., Dean, J.: Efficient estimation of word representations in vector space. In: 1st International Conference on Learning Representations, ICLR 2013 (2013)
19. Olah, C.: Understanding lstm networks. (2015). Online tutorial at: https://colah.github.io/posts/2015-08-Understanding-LSTMs/. Accessed 17 July 2020
20. Pennington, J., Socher, R., Manning, C.D.: GloVe: global vectors for word representation. In: Proceedings of EMNLP 2014 - Conference on Empirical Methods in Natural Language Processing, pp. 1532–1543 (2014)
21. Qin, Y., Song, D., Cheng, H., Cheng, W., Jiang, G., Cottrell, G.W.: A dual-stage attention-based recurrent neural network for time series prediction. In: IJCAI International Joint Conference on Artificial Intelligence, pp. 2627–2633 (2017)
22. Salinas, D., Flunkert, V., Gasthaus, J., Januschowski, T.: DeepAR: probabilistic forecasting with autoregressive recurrent networks. Int. J. Forecast. **36**, 1181–1191 (2020)
23. Zhang, G., Patuwo, B.E., Hu, M.Y.: Forecasting with artificial neural networks: the state of the art. Int. J. Forecast. **14**, 35–62 (1998)

Author Index

Printed in the United States
By Bookmasters